VEGETATION OF NEW JERSEY

VEGETATION OF NEW JERSEY

A Study of Landscape Diversity

Beryl Robichaud and
Murray F. Buell

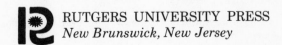
RUTGERS UNIVERSITY PRESS
New Brunswick, New Jersey

Library of Congress Cataloging in Publication Data

Robichaud, Beryl.
  Vegetation of New Jersey.

  Includes bibliographies.
    1. Botany—New Jersey—Ecology.   I.   Buell,
Murray Fife, 1905–   II. Title.
QK941.N5R6   581.9'749   72-4205
ISBN 0-8135-0745-6   Cloth
        0-8135-0795-2   Paper

Portions of this book appeared in *Diversity in Vegetation of New Jersey,* copyright © 1971, by Beryl Robichaud

The authors are grateful for permission to quote 29 lines from pages 3–4 of *The Pine Barrens,* by John McPhee, copyright © 1968, published by Farrar, Straus & Giroux, Inc.

Manufactured in the United States of America by Quinn & Boden Company, Inc., Rahway, New Jersey

Contents

Tables

Maps and Charts

Acknowledgments

This book has been made possible by the accumulation of information about New Jersey's natural landscape over a long period of time. Many individuals have contributed to this knowledge. Very detailed descriptions of the geology, soils, and forests of the State were recorded in bulletins and annual reports issued by the State Geologist in the years 1870 to 1917. Also available in published form are the late nineteenth- and early twentieth-century findings of well-known botanists including N. L. Britton, Witmer Stone, and John W. Harshberger.

Some of the more recent contributions to the information about the natural features of the State have been made by individuals associated with the Northeastern Forest Experiment Station, a unit of the U.S. Department of Agriculture; Dr. Silas Little was very helpful in making available such research data. Various members of the New Jersey State Department of Environmental Protection also were extremely helpful in giving us free access to unpublished photos and research data.

The chapter on New Jersey climate draws heavily on the contributions of Dr. E. R. Biel and the material covering the State's geology and soils in the publications of H. B. Kummel, Kemble Widmer, and J. C. F. Tedrow. In addition, special thanks are due Dr. Tedrow for his continued interest in the preparation of the book and for his critical reading of the chapter on the geology and soils. The work done by Louis Hand, one of the New Jersey's greatest field botanists, was of considerable value in preparing the descriptions of the State's vegetation. Also our gratitude is heartily extended to Dr. Barbara Palser who carefully read and criticized the first draft of the manuscript.

There is a long list of other people whose studies are very importantly involved in the establishment of a background for the book. These are members of a procession of graduate students at Rutgers University, candidates for masters and doctoral degrees, whose thesis research involved some aspect of New Jersey vegetation. During the last twenty-five years there have been more than sixty such students. A list of the individual names is too long to enumerate here but where major use of their published or unpublished results has been used in the text, reference is made to it. To all of them the authors express appreciation for their part in the knowledge and understanding of the various aspects of New Jersey's natural landscape.

Finally, special thanks must go to Dr. Paul G. Pearson of Rutgers University for his enthusiastic encouragement which stimulated the creation of this manuscript.

VEGETATION OF NEW JERSEY

Introduction

New Jersey, the fifth smallest of our fifty states, has a land area of only 7,509 square miles. Measured at its greatest length, from High Point near the northwest boundary to the southeastern tip of Cape May, the state extends 166 miles. At its greatest width from east to west the state is only fifty-seven miles wide; in midstate this distance narrows, and a line drawn from Trenton on the west to the Raritan Bay at the northeast would measure only thirty-two miles.

According to census figures, the 1970 population of New Jersey totaled 7,168,164, an average of 954 persons per square mile. This makes New Jersey the most densely populated of the fifty states. Perhaps more startling is the fact that New Jersey has more people per square mile than India, Japan, or the Netherlands, areas that usually are considered to be overcrowded.

Economically, New Jersey is highly industrialized and ranks seventh among all states in value of total industrial production. It is first of all states in the manufacture of chemicals, its leading industry. The state's other important industrial products include electrical machinery, food products, fabricated metal products, and transportation equipment. While the relative importance of agriculture has been declining sharply, 24 percent of the total land of New Jersey is still used for poultry and dairy farming or for the growing of vegetables, fruit, grain, hay, and ornamental plants. Of these products, eggs are the most important economi-

3

cally. In the mid-1960s the state ranked tenth in the number of eggs produced, second in the crop value of asparagus, and third in tomato cannery production.

Mining and quarrying operations, while less significant in the state's economy, show their imprint on the New Jersey landscape principally in the form of traprock quarries and sand and gravel pits.

To accommodate these industrial activities as well as the needs of its population, the state has built what is often described as "a beautiful network" of highways; it carries the densest flow of traffic in the world.

In spite of its relatively small land mass, its dense population, its advanced state of economic development, and its highly developed transportation facilities, New Jersey still has some natural vegetation. Even more surprising is the great variety in the vegetation—the landscape diversity—that now exists in the state. Although the main purpose of this work is to describe the vegetation of New Jersey in terms of its appearance and plant composition, it is equally important to explain why the vegetation is now what it is. The authors hope that a fuller appreciation and understanding of the present natural landscape will stimulate more interest in and concern about what we may leave behind for those who live after us. It is the people who have the ultimate responsibility for the use of our land resources and for the selection of actions that will have a significant impact on the future landscape of the state. Thus an informed public is necessary to ensure that wise decisions are made today to preserve a legacy for future generations.

To achieve these purposes, this book provides, first, a background for understanding why the vegetation of New Jersey is what it is today; second, a description of the present vegetation of the state; and finally, a look into the future of vegetation and man in New Jersey.

Throughout the text each plant is referred to by its common English name if one exists; the Appendix contains a two-way cross-reference between the common and scientific (Latin) names for each plant. No attempt is made to describe the botanical features of the hundreds of plants that are mentioned in the text. Instead, a list of references to plant identification is given in Appendix II. Also included in the Appendix is a guide to locations where the vegetation types of New Jersey can be seen.

REFERENCES AND SOURCE MATERIAL

New Jersey State Department of Labor and Industry, Division of Economic Development. 1964–1970. Facts about New Jersey, Facts and Facets of the Economy, A Guide to New Jersey Industrial Facts, Know Your State, County Data Sheets, and Estimates of Populations.

Part I
Understanding Vegetation

1

Understanding Vegetation:
Why the Vegetation of New Jersey
Is What It Is

Introduction

In this first chapter, definitions and fundamental concepts of vegetation are set forth as a basis for understanding what follows in the rest of the book. The basic concepts, relationships, and characteristics of vegetation that are described have general applications to other parts of the earth as well as to New Jersey.

Definitions

The word "vegetation" is used to refer to the total plant cover of a region, whatever its scale. Thus we may talk about the vegetation of one acre, of a whole state, or of the whole United States. The term "natural vegetation" in a strict sense suggests plant cover that has never been influenced by other than natural processes. However, because the impact of man has been so all-pervasive on our continent, the words "natural vegetation" most commonly are used today to refer to the plant cover that, while subject to disturbance by man's actions, grows and develops without man's purposeful intervention.

The plant cover of any one locale comprises different types

(or species) of plants that grow together. Simply described, a plant species is a particular type of plant that maintains its identity because it generally does not interbreed with other plant species. For example, a tree such as the white oak or sugar maple, a shrub such as the bayberry, and a flowering herb such as the common daisy that grows wild are all examples of different plant species. Each prevalent plant species is usually known by a common English name, but botanists identify each by a scientific name. The scientific label for a plant consists of two Latin words, the first classifying a plant by its genus category (like the surname of a person) and the second identifying each specific type of plant within a genus by its species name (like the first name of a person). For example, the scientific name for a common tree in New Jersey, the white oak, is *Quercus alba; Quercus* is the genus name used to identify all oak trees and *alba* is a word that designates the species of white oak as contrasted with the red oak, which is known as *Quercus rubra.*

The word "system" has been with us for many years. However, it has only recently been widely used by biologists, who are discovering living parallels to the solar system—a system composed of the sun, nine planets, and all other celestial bodies that orbit the sun. In such terms as "biological system," "solar system," and even "hi-fi system," the word "system" refers to a specific group of objects related to each other by mutual interactions. For example, the planet Earth is only one of many objects in the solar system, and it interacts with all the other planets and solar bodies. A speaker is just one component in a high-fidelity system, and to serve any purpose, it must interact with other parts of the system.

An ecosystem is a particular type of biological system in which plants, animals, and environmental factors are related to each other and affected by interactions. Thus, the vegetation of any region or area can and should be viewed as a component, and only one component, of a particular ecosystem. The word "ecosystem" is derived from "ecology," which is the study of the relationships among living organisms and their environment.

Vegetation as Part of a System — The Ecosystem

Since life began on this planet, living organisms have constantly interacted with each other and with environmental factors. These interactions have continually caused changes in the organisms themselves as well as in elements of their environment. Ecologists usually identify five ecosystem components which have interacting relationships — climatic elements, geologic and soil features, plants, animals, and man. Man is distinguished from other living organisms as an ecosystem component by his unique power and disposition to alter and even to exploit the other parts of the system.

Because this book is concerned with vegetation, interactions among the five ecosystem components are viewed primarily from the standpoint of a single component — plants. But the reader should keep in mind that important and interesting interrelationships exist among other components of the system — between climatic elements and soil-forming processes or between climatic elements and man.

Relationships between Climatic Elements and Vegetation

Perhaps most easily understood is the impact of climatic elements on plants; even good nurseries dealing with plants for the home gardener are careful to outline the water, temperature, and exposure requirements or tolerances of particular plant species. The three most important climatic factors affecting plant growth are temperature, precipitation, and light.

To plants as to all living organisms, temperature is important because it controls the rate of chemical reactions necessary for existence and growth. A plant's every physiological process involves temperature limits above and below which the process ceases and an optimum temperature at which it proceeds at a maximum rate. For example, little or no plant growth may take place during a hot summer day, and each plant species has an optimum temperature for germination of its seeds. With respect to temperature and to all other environmental factors, plant species differ in their tolerance to extremes and to the optimum for growth and for reproduction. Thus, while a species may flourish under the temperature, soil moisture, and other environmental conditions in one geographic region or even on one site within a region, it may be unable to reproduce successfully or even to exist in another region or site having a single environmental condition outside its tolerance.

Precipitation, whether in the form of rain, snow, or atmospheric humidity, is important to vegetation because some water is needed for all plant life processes. Water is required as a medium for chemical reactions and for the transport of mineral ions, both of which take place in every green plant. Lack of water often hinders plant growth, as can be observed with home plantings. As in the case of temperature, different plant species have different requirements for and tolerances of water.

The water that a plant uses comes mostly from soil moisture, which the plant absorbs through its roots. While much rainwater is lost to plants because of runoff and evaporation, precipitation provides most of the soil moisture used by natural vegetation.

The third important climatic factor affecting plants is light. Each species has an optimum light requirement because light is essential in the process of photosynthesis, by which a plant produces the food for its existence and growth. With too little or too much light, a plant may die. The seasonal differences in length of day that vary from one latitude to another are partly responsible for some of the geographical differences in vegetation.

Wind and atmospheric composition are among the other climatic factors that have an impact on vegetation. To some plants wind serves as the most important mechanism for the carrying of pollen from flower to flower and thus for aiding plant fertilization. Wind may also carry seeds from one site to another, making possible a wider dissemination of particular plants. But excessive winds can damage and kill plant buds and stems as well as reduce the amount of soil moisture available for plant development.

Only recently has attention focused on the effect of atmospheric pollutants on vegetation. Preliminary findings show that excessive pollutants in the air may weaken and even kill plants as well as humans. When a green plant takes in the carbon dioxide necessary for photosynthesis and the oxygen for its metabolic processes, it also will absorb gaseous or particulate forms of matter from industrial and household chimneys and other atmospheric contaminants. These may kill or weaken leaves or other parts of sensitive plants.

In all these ways climatic factors influence vegetation, while, to a lesser degree, vegetation has an impact on climate. As an example, the temperature on hot days and nights will be considerably lower in a forest than in a nearby city; wooded areas also act to decrease the force of heavy winds.

Relationships among Geologic and Soil Features and Vegetation

Geologic processes of mountain building, erosion, and glaciation sculpture our landforms by carving out ridges, hills, valleys, and even basins for lakes. Relief of land has an impact on vegetation because of its influence on drainage and exposure.

Even more important to vegetation, however, are the geologic processes which, interacting with climatic elements and living organisms, produce the soil in which plants grow. Soil, nothing more than disintegrated rock particles modified by plants, water, and air and combined with decomposed organic material, is the medium in which plants root and get their nourishment.

Soils differ in their water-holding capacities and in their supplies of mineral elements. One of the soil characteristics that affect the water-holding capacity is the soil texture—that is, the relative proportions of the sand, silt, and clay particles that make up soil. Water quickly drains through sand particles but adheres tightly to the smaller clay particles. The result is that while very sandy soils tend to be dry, those with a lot of clay often are waterlogged.

Soil substrates vary widely not only in water-holding capacity but also in the type and amount of the mineral elements most needed by plants—such as nitrogen, phosphorus, calcium, magnesium, potassium, and iron. Just as with the climatic elements, each plant species exhibits a specific range of need and tolerance for both water and particular nutrients as well as for other factors such as soil acidity. Thus a plant species that grows and reproduces successfully on one type of soil may not be able to exist under different conditions. With too little or too much of a specific element in the soil or too low or too high an acidity, as measured by a scale known as pH, a plant will die.

While soil characteristics have a great impact on plants, the plant cover, in turn, influences the process of soil formation. Soil organic material is derived from dead plant roots or from decomposed leaves and branches dropped to the soil surface as well as from the decayed remains of animals. In this way vegetation can significantly alter the characteristics of soil. The plants of the grassland leave a different imprint on soil than do trees of a forest, and even different forest types have distinctive impacts on soil.

Relationships between Animals and Vegetation

Ecosystem interactions between larger animals and plants are quite easy to observe even in areas settled by man. This is particularly true of the so-called "food-chain" relationships— which is the ecological term for the "who eats whom" relationships in nature. For example, leaves of trees that line residential streets in New Jersey may be eaten by canker worms or other foliage-eating insects which, in turn, may be eaten by birds such as the red-eyed vireo. And, if not alert, the vireo may be devoured by a hawk.

Plants are the initial food source in all food chains for, in spite of man's great accomplishments, the fact remains that only plants are able to convert inorganic materials of carbon dioxide and water to organic material called food.

Animals have both favorable and unfavorable effects on vegetation. Beneficial effects stem from the multitude of organisms that live in the soil. Some, like earthworms, improve the texture of the soil, and others aid in the decomposition of organic material. Animals also help plants to reproduce themselves. Flying insects and birds that carry pollen from flower to flower assist in fertilization, a necessary step in plant reproduction. Seeds develop from the fertilized flowers, and these may be embedded in berries or nuts or even barbed structures which easily attach themselves to animals. All three of these provide the means by which the seeds can be transported outside the range of the parent plant by a bird, squirrel, or other animal. Thus, animals help to ensure successful continuation of some plant species.

Unfortunately, not all relationships between animals and plants are beneficial to plants. Leaf-chewing insects can completely defoliate small plants and even whole trees. Such insects may, over a period of years, completely defoliate and kill an old established tree in a natural forest. Other insects, such as the elm bark beetle, may carry a lethal parasite from tree to tree, causing the spread of fatal tree disease. Too much grazing or browsing either by domestic animals such as cattle or by wild animals such as deer can alter severely the plant composition of natural vegetation.

Some unfavorable animal-plant relationships have been caused or amplified by actions of man. In the past, in the absence of man, balances were often maintained by natural processes. When the population of a particular insect or animal destructive to vegetation grew too large, it was often lowered rapidly by starvation (because the local food supply was used up) or by predator control. As will be discussed later, man sometimes upsets the balance of animal-plant relationships, with results disastrous to natural vegetation. This has happened in New Jersey as well as in other parts of the country and the world.

Relationships between Man and Vegetation

Only recently has there been general interest in and awareness of man's relation to the components of his ecosystem — climate, geologic and soil features, animals, and vegetation.

Regarding only the relationships of man and plants, several things are obvious. Without natural vegetation, man in his present form would not have developed at all. This is true for several reasons. First, oxygen in the form needed for human respiration did not always exist in the atmosphere. It developed only after the advent of plant life, as a product of plant photosynthesis. Next, man thus far has been unable to duplicate the photosynthetic activity of plants whereby nonliving materials are transformed into organic food. Thus man depends upon plants as the initial source for all his food as well as for much of the material used for his shelter and clothing. He may eat plants directly or indirectly — indirectly when he eats animals which, in turn, have depended on plants as the initial source of food in their food chain. While man is totally dependent on plant life for his existence, his actions from the beginning of his time on earth have been destructive to vegetation.

Many human activities have an impact on natural vegetation. Some of these are obvious and some are not. It is evident, for example, that natural vegetation is destroyed when land is used for agricultural crops, for dwellings, for industrial buildings, or for networks of highways. The accommodation of a larger and larger world population together with the trend toward more industrialization and urbanization means that each year less natural vegetation is left.

Less obvious are other actions of man that have an impact on vegetation. For example, whereas forests that are cut over lightly may show little change in species composition, lumbering activities that involve more extensive cutting of trees may change the composition of a particular forest so that it becomes quite different from what it was before it was first cut. Extensive grazing of domestic animals also causes changes in natural vegetation, and the consequences are an impoverishment in the plant life of grazed land. This has happened in many parts of the western United States as well as in Europe.

In still other ways man has had a drastic impact on natural vegetation. For example, by his past use of fire, man has greatly modified natural vegetation. It is believed that our ancestors learned to create fire hundreds of thousands of years ago. Having acquired this skill, man quickly became a great arsonist and

Figure 1-1 Picture at right shows multiple stems sprouting from the stump of a chestnut oak tree which previously had been cut back to the ground. Not all tree species have this ability to "resprout" after being cut or burned back.

caused fires throughout the land. Primitive man caused broad fire damage by abandoning campfires which then ignited surrounding vegetation, and he also probably started fires on purpose to facilitate hunting. As a result, extensive areas of the Old and New Worlds were burned many thousands of years ago.

As is evident in New Jersey, the Indians burned the land frequently and intentionally—to clear land for agriculture and settlement, to drive game in hunting, or to keep the forest open for travel. Frequent fires have greatly modified the natural vegetation in the state. This is so for several reasons. Some trees, by nature of their bark, are more insulated from heat than others and therefore less susceptible to fire damage. Seeds of some plants are destroyed by fire; the seeds of others germinate more quickly when exposed to the heat of fire. Also, some trees and shrubs are able to sprout quickly after fires while others have no resprouting capability (Figure 1-1). New Jersey offers excellent examples of the effect of fire on vegetation, and these are described in Part IV of this book.

Although man's use of fire no longer has the impact on vegetation that it once did, a new destructive force has become important recently—pollution. The ultimate results of man's actions in polluting the atmosphere, the soils, and the waters are not yet clear. However, it is known that excessive amounts of atomic radiation and certain chemical wastes such as sulphur dioxide may modify natural vegetation or destroy it completely. Water pollution often leads to the excessive growth of undesirable aquatic plants. Another relatively new source of serious damage to vegetation stems from the indiscriminate use of pesticide chemicals. These have been designed to control destructive insects or plant fungus diseases, but the results have had a boomerang effect in many cases and have sometimes caused greater rather than less insect or disease damage to plants because of the interference with the natural balance inherent in nature's food chains.

Ecosystem Characteristics

All ecosystems, those in New Jersey as well as those in other parts of the world, have two characteristics common to all living biological systems. These characteristics, depicted graphically in Figure 1-2, are as follows:

1. Each ecosystem is continually changing in time. In any one place on earth, the interactions among climatic elements, geologic and soil-forming processes, animals, plant populations, and humans are not the same today as they were yesterday. In all ecosystems each of these components is changing continually although at differing rates from time to time. Thus the natural environment may be said to be "dynamic," meaning that climate, landforms and soils, animals, plants, and even man himself have

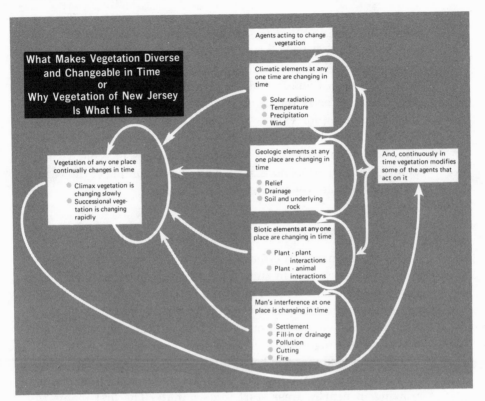

Figure 1-2 Schematic diagram showing ecosystem relationships which cause change (succession) in natural vegetation. Because these relationships differ from one place to another, there is variety (diversity) in natural vegetation.

changed each day in the past and will continue to do so in the future.

2. Each ecosystem is distinctive—that is, the conditions in any one place, such as one small area within a state, are not exactly the same as in any other area. This is because particular combinations of climate, geologic history, and interactions of animals, plants, and man found at any one place will not be exactly duplicated in any other place. Thus it is said that ecosystems and their components vary in space.

The two characteristics noted above apply to each component

in an ecosystem as well as to the total ecosystem itself. Thus it should be expected that the plants growing on any one site will change in time, thereby resulting in a continual natural change of vegetation. Ecologists call this process of change in vegetation with time succession. In addition, the plants growing on any one site are not exactly the same as those growing on another site, thereby making for variety in vegetation. This variety is called diversity by ecologists. These two notions—succession and diversity—are among the more important concepts of ecology, and knowledge of them is necessary to an understanding of the vegetation of New Jersey or of any other region.

Succession—The Change of Vegetation in Time

Without man's interference the vegetation of any particular site does change in time. This can be easily verified by observations of real events. For example, Dr. Gily Bard, an ecologist who studied land near New Brunswick, New Jersey, that once had been cultivated for agriculture but then left idle, found that the following natural changes took place (Figure 1-3).

When a field that had been cultivated is left abandoned, seeds that have been dormant in the soil or that have come from plants growing in nearby areas start to germinate. The first plants to grow successfully are those with short lives. Of these, ragweed, the annual plant whose pollen causes the hay fever allergy, is the most abundant. Annual plants are so named because they live only one year and then die. Also, common invaders in the first year are the wild radish and wintercress; the mass presence of the latter often can be seen in May and June as a blanket of yellow blooms on an abandoned field in New Jersey.

By the second year other plants develop in the field, such as the goldenrods and asters. These are perennial plants that have a life span of more than two years, and their continued spread over several years results in their gaining almost complete dominance after a period of time (Figure 1-3). Other herbs start to grow in the field, one of the most abundant being a high-growing wild grass known as little bluestem. Less common but conspicuous for their flowers are the herbs known as Queen Anne's lace, common mullein, daisy, and black-eyed susan.

For as many as ten years, the mixture of low-growing herbaceous plants gives the appearance of a natural flower garden. Then the scene changes, for seeds of woody shrubs and trees successfully germinate in the field and develop into plants that grow taller than the herbs. In mid-New Jersey the first woody plant to show up conspicuously in an abandoned field is the red cedar tree, a narrow, pyramid-shaped tree with small evergreen leaves that remain on the tree the year round. As these grow to a maximum height of about 30 feet, they are joined by seedlings of other trees such as the red maple, wild black cherry, and sassafras. In addition, lower-growing shrubs such as bayberry occupy places in the field, and scattered thickets of shrubs and young trees interwoven by vines such as poison ivy become conspicuous. In twenty to thirty years, either the little bluestem grass crowds out the showy flowering herbs that once grew so abundantly in the field or the trees shade them out. Finally, in fifty to sixty years' time, the piece of ground that had been an abandoned field will be changed into an open woodland of fairly tall trees (Figure 1-3).

Some of the reasons for natural succession become obvious when one remembers that all ecosystem components change in time. For example, as taller plants occupy a site, less light falls on the surface of the soil. This makes the environment less favorable for plants which have a high requirement for light. Some plants just cannot reproduce or even develop successfully in shady conditions and thus will die off as trees grow up around them. On the other hand, many trees that can develop under shady conditions cannot seed themselves in an abandoned field until it has first been occupied by herbaceous plants. Changes in vegetation also are accompanied by alterations in soil conditions and in climatic conditions close to the ground (the microclimate). There is a constant struggle or competition among plants for resources of water, soil nutrients, light, or whatever may be in short supply. Slight changes in environmental conditions can favor one or more species over others.

6 Years after abandonment
Field is covered with herbs,
mostly goldenrod and asters.

12 Years after abandonment
Red cedar trees have invaded
the field and a wild grass
(little bluestem) has replaced
the pioneer herbs.

Figure 1-3 The process of natural change (succession) in fields left
abandoned, Franklin Township in Somerset County.

About 30 years after abandonment Red cedar trees still are the most dominant plants; gray dogwood and blackberry shrubs are growing to the left of the cedars. Bayberry shrubs and sumac and sassafras trees also are common in the field.

About 50 years after abandonment A woodland now occupies the field. Red maple, black cherry, sassafras, and pin oak trees are abundant and will shade out the lower-growing red cedar trees and the shrubs.

If changes occur relatively rapidly in the plant growth of a particular site, the vegetation is referred to as "successional" vegetation. On the other hand, if and when a point is reached at which change becomes extremely slow, the plant growth may be called a "climax" type of vegetation. For example, the extensive forests of spruce and fir trees that cover a great part of the undisturbed land of Canada are considered climax vegetation.

Man's actions frequently interfere with the natural course of succession in vegetation; for example, he may destroy a forest of climax vegetation and, in so doing, start a new series of successional vegetation. In examining vegetation, it is important that we distinguish between natural and man-made changes. This is necessary because in some cases man has disturbed the vegetation so greatly that it will take many years to revert to its natural stages of development.

Diversity – The Change of Vegetation in Space

The distinctiveness or individuality of each ecosystem stems from the spatial differences that exist from place to place in climatic factors, geologic and soil features, and interrelationships of animals, plants, and man. This is what gives us variety or diversity in vegetation. Because each plant species has different requirements and different ranges of tolerance to various environmental factors, we should expect some variety in plant composition as changes occur in environmental conditions. Obviously, then, the degree of variation in vegetation from one geographical location to another (or even from one site to another within the same geographic area) will depend in great part on the relative differences in climate, geologic and soil features, animal-plant interrelationships, and the extent of man's interference. More will be said about this in Chapter 6.

Summary

The concept of vegetation as one, and only one, component of the ecosystem explains the two basic characteristics of all natural vegetation—its change in time (succession) and its change in space (diversity). The chapters in Part II deal with each of the ecosystem components individually—geologic and soil features, climate, man's actions, and other biological influences (plant-plant and animal-plant relationships) as each has affected the vegetation of New Jersey.

References and Source Material

Bard, Gily E. 1952. Secondary Succession on the Piedmont of New Jersey. Ecological Monographs 22: 195–215.

Dansereau, Pierre. 1957. Biogeography: An Ecological Perspective. Ronald Press, New York.

Odum, Eugene. 1966. Ecology. Holt, Rinehart, and Winston, New York.

Stewart, Omer. 1956. Fire as the First Great Force Employed by Man. *In* Man's Role in Changing the Face of the Earth. University of Chicago Press, Chicago. Pp. 115–133.

Treshow, Michael. 1970. Environment and Plant Response. McGraw-Hill, Inc., New York.

Part II

Influences on Vegetation of New Jersey

2

Geologic and Soil Features
of New Jersey

Introduction

Small as it is, New Jersey still bears the imprint of highly
varied geologic processes operating throughout hundreds of
millions of years. Some of the results are easily recognized today
by the diverse relief, varied parent rock and other surface material, and complex drainage patterns found in the state. It is also
primarily these three geologic features that both influence and
interact with vegetation. Relief is important because higher elevations have lower temperatures, greater exposure to winds, and
usually thinner, less fertile soil—and steep slopes have greater
water runoff. Because soils are formed primarily from the disintegration of parent rock or surface deposit material, they can
vary widely in qualities of texture, water-holding capacity, and
nutrients—all important to plants. Finally, drainage patterns
including water table levels, coastal tidal areas, and inland
river systems all influence vegetation.

In this chapter the ecosystem relationship of geologic and
soil features to vegetation in New Jersey is examined. The chapter starts with a simple explanation of the geologic time scale and
a definition of the present geologic divisions of New Jersey. This
is followed by a brief outline of the state's geologic history and a

description of each geologic region from the standpoint of characteristics that have an impact on vegetation.

The Geologic Time Scale

Geologists date past events on a scale in which time is divided into major divisions called eras and each era, in turn, is subdivided into smaller units called periods. While the earth is now thought to be many billions of years old, it still is not possible to describe accurately geological events dating back to earliest times. Instead, for purposes of geological and historical biological descriptions, many scientists go back only 510 million years to begin their fine breakdown of history. According to the U.S. Geological Survey, anything that happened before 510 million years ago is said to be in Precambrian time. Everything more recent than that is classified in three eras. The *Paleozoic era,* the era of "ancient life," is the time that stretches from approximately 510 million to 180 million years ago. The *Mesozoic era,* the era of "middle life," starts at about 180 million years ago and continues to about 63 million years ago. And the *Cenozoic era,* the era of "new life," starts about 63 million years ago and includes the time in which we live. Rather than burden the reader with full details of the division of eras into periods and still finer classifications, only those specific periods most important to geologic development in New Jersey will be identified and discussed herein. The references at the end of the chapter include publications containing more complete descriptions of the geologic history of New Jersey.

Physiographic Land Regions in New Jersey

As a reference framework for explaining the geologic history that follows, it is necessary to   delineate the present physiographic land divisions in New Jersey. The state is divided into five very different physiographic regions known as:

Ridge and Valley Section
Highlands Section

Piedmont Section
Inner Coastal Plain Section
Outer Coastal Plain Section

Figure 2-1 shows the boundaries of the five regions in relation to a county map of New Jersey and to a relief map of the state.

Geologic History of New Jersey

Because we live in such a relatively tiny time span compared with the earth's history, it may be difficult for some of us to recognize that geologic processes, like the other components of the ecosystem, are operating continually. Mountain building, erosion, glaciation, and wave action on coastal dunes, important geologic processes in New Jersey's past, are still sculpturing landforms. Change is taking place continuously. And as illustrated in Figure 1-2, geologic and soil features, climatic elements, and living organisms, while changing continuously in time, are all interacting with each other as well. We will look back briefly into the geologic history of New Jersey to identify those events that have had the greatest impact on the landscape of the state.

Earliest History—the Precambrian Era
(More than 510 Million Years Ago)

The rocks making up the Highlands were first formed more than 510 million years ago. Some were the result of lava flows or volcanic eruptions; others were igneous or sedimentary rocks, the latter formed while the land was covered by seas. After their original formation, nearly all Precambrian rocks were subject to intense heating and pressures and became what are known as metamorphic rocks.

Early History—the Paleozoic Era
(about 510 to 180 Million Years Ago)

At various intervals during the Paleozoic era, ocean waters, coming from the southeast, covered New Jersey for millions of

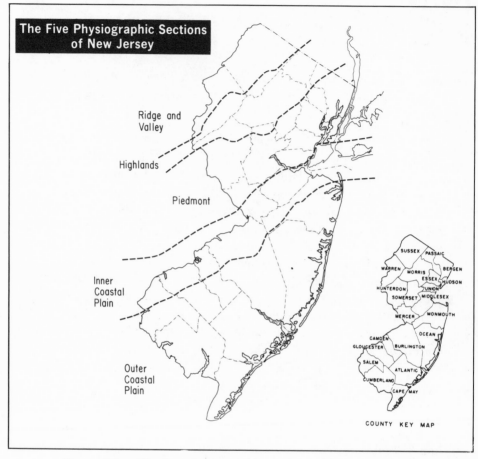

The Five Physiographic Sections
of New Jersey

Ridge and
Valley

Highlands

Piedmont

Inner
Coastal
Plain

Outer
Coastal
Plain

SUSSEX
PASSAIC
WARREN
MORRIS
ESSEX
BERGEN
HUDSON
HUNTERDON
UNION
SOMERSET
MIDDLESEX
MERCER
MONMOUTH
CAMDEN
OCEAN
GLOUCESTER
BURLINGTON
SALEM
ATLANTIC
CUMBERLAND
CAPE MAY

COUNTY KEY MAP

Figure 2-1a   The five physiographic sections of New Jersey related to
county boundaries.

years. The present Ridge and Valley section of New Jersey shows
the imprint of the Paleozoic era most prominently in its sedimentary rock formations. Layer upon layer of sediments accumulated on the ocean bottom, and when these were uplifted, extensive limestone and sandstone rock formations became exposed
land. Geologists identify land areas previously covered by oceans
through the fossils of past marine animal and plant life that are
found embedded in the rock formations. The sedimentary deposits

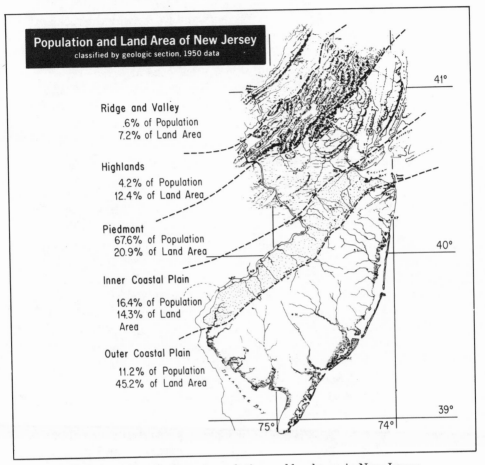

Population and Land Area of New Jersey
classified by geologic section, 1950 data

Ridge and Valley
.6% of Population
7.2% of Land Area

Highlands
4.2% of Population
12.4% of Land Area

Piedmont
67.6% of Population
20.9% of Land Area

Inner Coastal Plain
16.4% of Population
14.3% of Land
Area

Outer Coastal Plain
11.2% of Population
45.2% of Land Area

41°
40°
39°
75°
74°

Figure 2-1b   The distribution of population and land area in New Jersey by the five physiographic sections of the state. Source of map data: John Brush. The Population of New Jersey. Rutgers University Press, New Brunswick, New Jersey, 1956.

were subsequently folded, eroded, and refolded during the times when the land of Eastern North America was uplifted and folded, forming the Appalachian Mountains.

The sediments deposited when the seas cover the land are always subject to subsequent erosion. This erosion occurs because, when the land is uplifted and the seas withdraw, streams drain the land for millions of years and wear away the rocks. As rocks have different degrees of hardness, the process of erosion produces relief in landforms—that is, ridges and valleys. Harder rocks form the ridges while the softer rocks are eroded down to valleys or plains. Uplift of land and subsequent erosion is a continuing process. It happened in the past, it is happening today, and it will continue as long as the earth exists.

Middle-Age History—the Mesozoic Era *(180 to 63 Million Years Ago)*

Only two of the three subdivisions of the Mesozoic era are important in the geology of New Jersey. The first is the Triassic period, which started about 180 million years ago and lasted about 30 million years. During this period, the red sandstone and shales that presently form the surface of the Piedmont were deposited. This happened when the land again was covered by the seas. In some places the sandstone sediments are 12,000 or more feet thick. During the Triassic period, the climate alternated between wet weather cycles and long, very dry spells in which large salt lakes formed. The imprint of these fluctuations of weather is left in the sedimentary formations.

After most of the sandstone was deposited, volcanic (or igneous) activity with lava flows occurred in the Piedmont section of New Jersey. These lava flows resulted in the formation of the present Watchung Mountains, the Palisades, and other diabase and basaltic ridges that rise about the Piedmont plain, such as the Sourland Mountains and Cushetunk Mountain. During the Mesozoic era, as before and after, uplift of whole areas of land continued and was followed by millions of years of erosion by inland rivers. The erosional processes etched out the present relief of the Piedmont, leaving the harder volcanic rocks to form the ridges and hills and the less resistant shale and sandstone to form the lowlands.

The Triassic period was followed by the Jurassic, which was relatively unimportant in the geologic history of the state. Then followed the Cretaceous period, which was very important in New

Figure 2-2 It is the process of erosion that has carved out the present relief of New Jersey. Diagram shows a cross section of the state from its northwest corner to Barnegat Bay in the southeast.

Jersey geologic history. In this period (125 to 63 million years ago), the seas again encroached upon the land of southern New Jersey and withdrew, only to return again and again. This happened many times, and each time that the sea covered the land, sedimentary deposits composed mainly of clays, silts, sands, and gravels were left. Each time the sea withdrew, erosion proceeded and streams carried away some or all of the sediments, depending upon the hardness of the material. Incidentally, it was from the Triassic to the Cretaceous period that large, duck-billed, plant-eating dinosaurs roamed the land of New Jersey.

Recent History — the Cenozoic Era (from 63 Million Years Ago to and Including the Present)

The imprint of the Cenozoic era is also visible in New Jersey today. This era is divided into two time intervals, the first of which is called the Tertiary. Tertiary time started 63 million years ago and continued to 1 million years ago. Many times during this interval the seas invaded that part of southern New Jersey known as the Outer Coastal Plain. With each invasion, clays, silts, sands, and gravels were deposited and subsequent erosion carried some of the deposits back to the sea. While these deposition and erosion processes were occurring on the Outer Coastal Plain, the land in other parts of New Jersey was being uplifted, eroded, and sculptured into ridges and valleys. Since the last time the land was uplifted, streams have carved out the present valleys and ridges in the northern part of the state.

A very important part of New Jersey's geologic history took place in Quaternary time, the interval in which we live. Quaternary time usually is dated from 1 million years ago and includes the period of the Pleistocene glaciation. Since then huge ice masses that had accumulated in Labrador and the Hudson Bay region have moved southward in North America. There have been four major ice advances which are referred to as the Nebraskan, Kansan, Illinoian, and Wisconsin ice ages, the Wisconsin advance being the most recent. In addition, there have been minor but locally important ice movements. Between the successive ice advances extended periods of warmer weather have occurred, causing snow to melt and sea levels to rise. During these interglacial periods, patches of yellow sand and gravel were deposited on parts of the Inner and Outer Coastal Plains.

At least three of the four ice advances — the Kansan, Illinoian, and Wisconsin — reached northern New Jersey. As the ice sheets moved across New Jersey, they smoothed out rock formations and cut scratches and grooves in underlying bedrock, some of which remain. All existing vegetation was destroyed except that which lived in the ice sheet. Rock fragments were picked up and carried south with the ice sheets. When the ice melted, the rocks carried from the north were left in the south. Glacial streams also transported smaller rock debris from the north to more southern areas having different underlying parent rock formations. The glacial deposits were thickest at the southernmost end of the ice sheets, and these are known as the "terminal moraines." The terminal moraine of the Wisconsin glacial ice sheet stretched in an irregular belt almost 1 mile wide from Perth Amboy at the east, through the Plainfield area, to Belvidere in the western part of the state (Figure 2-3). The terminal moraine as it still exists today is a heterogeneous mixture of rocks and gravel, as deep as 300 feet near Dover.

In addition to changing landscape by its sculpturing force and deposits, glaciation left an impact on stream drainage pat-

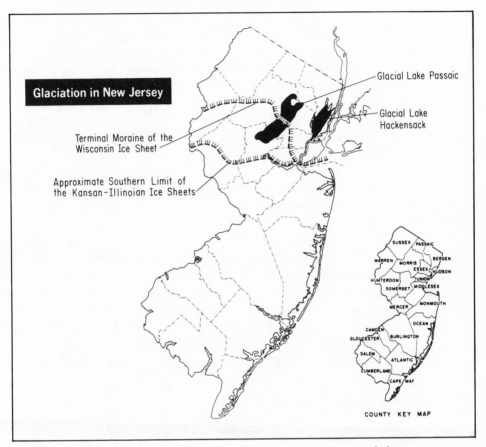

Figure 2-3 The locations of glacial terminal moraines and former glacial lakes in New Jersey. Map redrawn from Kemble Widmer. The Geology and Geography of New Jersey. D. Van Nostrand Co., 1964.

terns. In New Jersey as in other areas, natural lakes were formed because glaciers scoured out depressions in rock beds and glacial deposits blocked previous stream drainage channels. The state's largest lakes, Hopatcong, Greenwood, and Budd lakes, are glacial-formed lakes, and most of New Jersey's natural lakes are located in just that part of the state that was covered by the ice sheets. In addition to the lakes just mentioned, two other very large lakes existed in glacial times—Glacial Lake Passaic and

Glacial Lake Hackensack (Figure 2-3). These lakes came into being because glacial deposits blocked and changed the drainage patterns of the Passaic and Hackensack rivers. While many of the glacial lakes have since drained, their mark is left in distinctive marsh or swamp land areas such as the Great Swamp near Chatham and the Troy Meadows near Whippany, both of which are remnants of Glacial Lake Passaic.

Glaciers, by their presence or absence, affect sea levels, which are lowest when glaciation is at its maximum and highest when glacial ice is at a minimum. This explains why the sea levels on the New Jersey coast are now rising as the glacial ice continues to melt. It also explains the presence of certain deposits left on the Coastal Plain during the interglacial periods.

After the glacial ice sheet disappeared about 12,000 to 15,-000 years ago, some of the land that had been depressed by the weight of the ice sheet was uplifted. This was true of the New Jersey coastal area. Since the uplift, marine erosion and deposition processes, primarily in the forces of wave and tidal action, have been at work sculpturing coastal land. Nowhere in New Jersey are the processes of geologic change more apparent today than on its coast. The results are described in a later chapter.

Present Landforms in New Jersey

The geologic history just described has left in New Jersey a naturally diverse and complex landscape. Within its small area, there is considerable variation in landforms, in surface and parent rock materials, and in the soils derived from them. All these, combined with complex drainage patterns, have an impact on vegetation. The highlights of these features in each of the five physiographic regions are illustrated in Figures 2-4, 2-5, and 2-6 and described below. Serving as supplements to the text descriptions are two maps in Figures 2-7 and 2-8 at the end of the chapter. The first shows in simplified form the geologic bedrock of New Jersey and the second a classification of New Jersey soils as made by Dr. John Tedrow, soil scientist at the College of Agriculture and Environmental Science of Rutgers University.

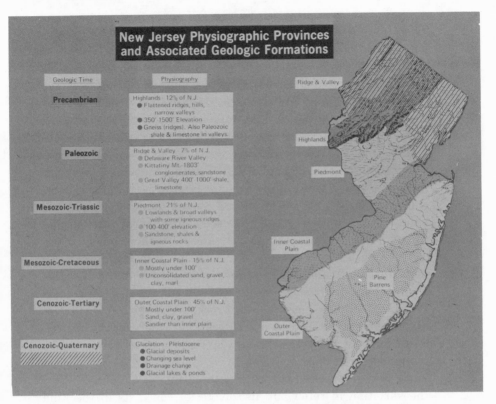

Figure 2-4   The five physiographic sections of New Jersey, the geologic age and bedrock of each, and its relief features.

Ridge and Valley Section

The Ridge and Valley section of New Jersey (also called the Valley and Ridge section or simply the Valley section) is part of a larger Ridge and Valley geologic province that extends from Canada to the southern United States as a narrow belt of ridges and interconnecting valleys having a northeast-southwest orientation. These same features are represented in New Jersey. In the far west of the state is the valley of the Delaware River, which marks the western border of the state. At its northern end in New Jersey the Delaware Valley is about 500 feet above sea level.

Kittatinny Mountain separates the Delaware from another and broader valley in the east; it is a flat-topped ridge varying in width from 1 to 5 miles. Kittatinny reaches a maximum height of 1,803 feet in the northwest, at High Point, the highest altitude in the state. To the east of the mountain is Kittatinny Valley, which is part of the eastern United States Great Valley. This valley section in New Jersey is 40 miles long and about 12 miles wide. It has a broad, undulating contour varying in elevation from about 400 to 1,000 feet above sea level.

Altogether the Ridge and Valley section contains 635 square miles or about 7 percent of the total land in New Jersey. It occupies a large part of Warren and Sussex counties. Ridges and valleys occur in this section because different parent rock formations underlie the ridges and the valleys. As mentioned earlier, softer rocks such as limestone and shale erode faster than the more resistant sandstone and conglomerates. The lowest valley levels occur wherever limestone underlies the surface; the areas of shale, a slightly more resistant rock, are about 200 to 400 feet higher than the limestone, and ridges occur wherever the bedrock is of material more resistant to erosion, such as sandstone or conglomerate rock.

The differences in parent rock material not only account for the variation in relief but also make for contrasts in the kind and amount of soil coverage. In general, the soil covering the Kittatinny and other ridges in this section is poor in quality from the standpoint of vegetation. The soil layer is thin on the ridges, with bedrock exposed in many places. Also, the ridge soil tends to be very acid and of low fertility and, often, very stony.

In contrast, the soils in the valleys, derived from limestone and shale that were covered by glacial till, are for the most part deeper and more fertile, and are well drained. Peat or large muck deposits (thick layers of organic material) may occur where shallow glacial lakes once existed. These were later invaded by vegetation, the dead remains of which accumulated as peat or muck.

Highlands

Southeast of the Ridge and Valley section is the Highlands, a part of a larger geologic province called the New England Uplands, which includes the Green Mountains of Vermont. In New Jersey the Highlands have an area of about 900 square miles, or 12 percent of the total land area in the state. As shown in Figure 2-4, this section is broader at the north, where it is about 20 miles wide; at its southern end bordering the Delaware River Valley, it is only 10 miles wide. The Highlands also has parallel ridges and valleys, but these differ from the Ridge and Valley section in the type of parent rock underlying the surface. Also, the ridges are more massive and generally much broader while the valleys are narrower and have steeper slopes. Frequent rock outcroppings occur. Glacially formed lakes, such as Lake Hopatcong and Green Pond, contrast with adjacent ridges to make the Highlands a very scenic area of New Jersey.

The elevation in the Highlands averages about 1,000 feet above sea level, with a ridge maximum in the northwest of about 1,500 feet. The southern part of the Highlands shows a more gentle contour, with the valleys reaching a low of 350 feet. While the oldest rocks in the state are in the Highlands, the ridges in this section have resisted erosion because they are made primarily of gneiss, a very hard rock material. The Highland valleys are of softer limestone or shale. The soil derived from the gneissic parent rock differs within the Highlands according to whether the underlying bedrock was covered by older glacial drift. More important to vegetation, however, are the variations in relief, soil drainage, and stoniness that occur within the area.

Piedmont of New Jersey

East and south of the Highlands is the section called the Piedmont or the Triassic lowlands. The New Jersey Piedmont belongs to a larger belt of rock formation that extends almost 1,000 miles from the Hudson River southward through New Jersey and Pennsylvania where it is separated from a similar more southerly formation that extends through Maryland into

Figure 2-5 The landforms of northern New Jersey.

Photos on opposite page

Above, in the Ridge and Valley section. The view from Kittatinny ridge looking eastward toward the Kittatinny valley; in the background the hills of the Highlands.

Below left, in the Highlands section — Laurel Lake in Wawayanda State Park.

Below right, in the Highlands section — the Great Meadows, the site of a former glacial lake.

Photo below, in the Piedmont section looking eastward toward the Watchung Ridges.

Virginia. Detached areas of the same formation also occur northward in Connecticut and Massachusetts.

The Piedmont of New Jersey occupies about 21 percent of the land area and is composed mostly of shale, sandstone, and argillite formations that typically are Indian-red or brownish-red in color. These formations are less resistant to erosion than the adjacent Highland gneissic rock and so, in comparison to the Highlands, the Piedmont is, in fact, a lowland. The Piedmont section in New Jersey slopes gently southeastward from about 400 feet above sea level at its northwestern margin to an elevation less than 100 feet at its southern margin bordering the Delaware and to sea level at Newark Bay. Though flat in some areas, the Piedmont contour is slightly rolling, with mostly gentle slopes, but in some areas rivers have cut rather steep-sided valleys.

Interestingly, on the Piedmont there are several ridge formations that tower over the adjacent lowlands – the three Watchung Mountains (850, 650, and 350 feet high), Cushetunk Mountain, the Sourlands, and the Palisades. These ridges are made of intrusive or extrusive lava material known as diabase and basaltic rocks, both of which are much harder than the shale and sandstone of the Piedmont. While the diabase and basalt have resisted erosion, the less resistant shale and sandstone have been worn down, resulting in the lower elevations.

Differences in the rock formations, combined with the fact that the glacial deposits of varying age covered only part of the Piedmont, have resulted in a variety of soil types within the area. However, these variations appear to be less important to vegetation than the variation of soil water drainage.

Inner and Outer Coastal Plains Sections

The Coastal Plains area is the most easterly and southern part of New Jersey and comprises about 60 percent of the total land area of the state. It belongs to a larger geologic province of the eastern United States that extends northward through Long Island to Cape Cod and southwestward along the coast into Mexico.

Although both the Inner and Outer sections of the Coastal Plain in New Jersey have their origin in depositions of clays, silts, sands, and gravels, the two areas are distinctive particularly with respect to soils. The land identified as the Inner Coastal Plain represents sedimentary deposits made in the Cretaceous period that were covered for the most part with later deposits made in interglacial Pleistocene time. The Outer Coastal Plain, on the other hand, consists of sedimentary deposits dating from Tertiary time but with overlying patches of sand and gravel deposits that also date back to interglacial time.

The Inner Coastal Plain is separated from the Outer Coastal Plain by a belt of hills which extends in a southwesterly direction from the Atlantic Highlands (the Highlands of the Navesink) overlooking Raritan Bay to the Delaware River lowlands in the southwest. The hills are remnants of a particular landform called a cuesta. On the lower elevations of the Coastal Plain the clays, silts, sands, and gravels, for the most part, are unconsolidated; that is, the particles are not cemented together as in a sandstone rock. On the cuestas, however, some of the sands and gravels are cemented together to form a rocklike cap on the hills. The larger cuestas include the Beacon Hill formation, which reaches a height of 373 feet; Arney's Mount, 230 feet; Mount Holly, 183 feet; and Mount Laurel, 173 feet. These larger cuestas contrast with most of the rest of the Coastal Plain area, which is less than 100 feet in elevation.

The Inner Coastal Plain lies to the west of the cuestas, and water drains westward to the Delaware River or north to the Raritan Bay. On the Outer Coastal Plain, which lies to the east of the cuestas, water mostly drains eastward on a gentler slope to the ocean or south and southeast to the Delaware Bay. (The Rancocas Creek is one exception; it flows westward from its source in the Pine Barrens to drain into the Delaware River.) Thus the whole surface of the Coastal Plain may be pictured as rising gradually from sea level on the east, west, and south to elevations as high as nearly 400 feet where the Inner and Outer Coastal Plains join at the cuestas.

Just as the continual process of differential erosion determined the relief in the other three physiographic sections of New

Above, on the Inner Coastal Plain near Mt. Holly looking eastward toward Arney's Mount, a cuesta hill formation, that marks the separation here between the Inner and Outer Coastal Plains.

Figure 2-6   The landforms of the Coastal Plains in southern New Jersey.

At left, the sandy Pine Barrens of the New Jersey Outer Coastal Plain. *Below,* sand dunes on the offshore islands on the New Jersey coast.

Geologic Bedrock of New Jersey

Figure 2-7   Geologic map of New Jersey showing major bedrock formations. Map was copied from one prepared by J. C. F. Tedrow (1962). The
source of the map key also was from Tedrow (1962) with additional data
extracted from the official geologic map of New Jersey prepared by J. V.
Lewis and H. B. Kummel in 1910–1912 and revised by Kummel and
others.

Key to Geologic Map

(Age of bedrock is shown in parentheses)

Sandstone and shale sediments with some limestone, gray in color (Devonian).

High Falls formation of hard red sandstone and soft red shale (Silurian).

Shawangunk Conglomerate, hard, dark gray (Silurian).

Martinsburg Shale, dark gray slaty shale (Ordovician).

Kittatinny Limestone, gray (Cambro-Ordovician).

Gneiss, granite, gabbro and metamorphic rocks (Precambrian).

Green Pond and Skunnemunk Conglomerates—red; Bellvale Sandstone—gray, and Pequanac Shale—gray (Devonian and Silurian).

Basalt and diabase—igneous rocks called "traprock" (Triassic).

Stockton formation of gray, feldspathic sandstone, conglomerate, and red shale (Triassic).

Lockatong formation of hard, reddish to blue-gray argillite and argillitic shale (Triassic).

Brunswick formation of soft red shale and sandstone but with conglomerate beds with some limestone pebbles (Triassic).

Brunswick formation of soft red shale and sandstone (Triassic).

Sediments of unconsolidated sand, silt, and clay, many high in glauconite (Cretaceous).

Sediments of reddish sand, colored and consolidated in places by iron oxide (Cretaceous).

Sediments of unconsolidated, yellow to gray, quartzose sands with a few clay lenses (Tertiary).

49

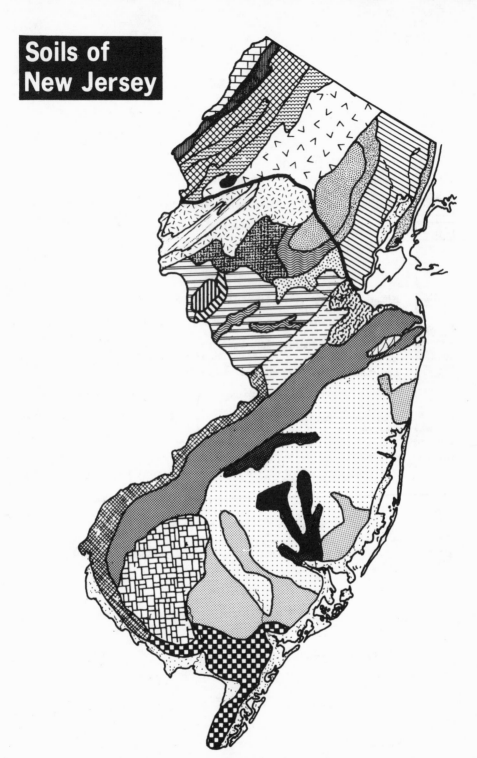

Figure 2-8   Classification of soil types in New Jersey. The soil classification map is copied from J. C. F. Tedrow (1962). The text for the map key was extracted from Dr. Tedrow's (1962) description of soils.

Wallpack soils are generally deep and well drained but stony and shallow on steep slopes.

Cattaraugus-Swartswood undifferentiated soils are mostly acid and stony and shallow on steep slopes.

Nassau-Dutchess-Cossayuna undifferentiated soils are well drained and shallow in some places.

Palmyra-Squires undifferentiated soils are generally loamy, well drained, moderately acid at the surface with carbonate at greater depths.

Rockaway soils are acid and mostly well drained with stony conditions on steep terrain.

Wethersfield soils are generally loamy with good drainage.

Holyoke soils on traprock are mildly acid, stony, and sometimes shallow.

Muck, organic accumulation, is acid in reaction.

Whippany soils, formed on the site of Glacial Lake Passaic, are mostly poorly-drained silts.

Dunellen soils are generally sandy, acid, deep, and well drained.

Annandale are deep, well-drained soils but often stony on steep terrain.

Washington soils are among the best in North Jersey – mildly acid, friable, deep, and well-drained silt loam.

Norton are deep, well-drained, loamy soils.

Penn soils are shallow, well drained and loamy but sometimes slightly droughty.

Montalto are deep, well-drained, mildly acid soils but on mostly steep and stony traprock slopes.

Lansdale soils are deep, well-drained, moderately acid soils on rolling to steep sandstone terrain.

Croton soils are wet, compact, acid, and silty.

Sassafras are among the best soils in the state having good aeration, drainage, and loose sandy conditions.

Sassafras-Hammonton Phase soils are distinguished by their lower nutrient retention capacity and droughty character because of unusual sandiness.

Sassafras-Cape May Phase soils are well drained in upper horizons but with a water table in the lower part of the soil.

Sassafras-Keyport undifferentiated soils are usually sandy and fairly well drained but of slow drainage where high in silt and clay.

Aura soils are formed on well-drained acid, loose sands and silts but a hard subsoil condition occurs at a depth of about two feet.

Greenwich soils are well drained and high in fine sand and silts.

Freehold-Collington undifferentiated soils are excellent – deep, well drained and contain small amounts of glauconite.

Lakewood, the soils in the drier Pine Barrens, are very sandy, acid and highly leached.

St. Johns soils occupy the very acid, sandy wet sites in the Pine Barrens.

Colts Neck soils, formed in the reddish sandy Coastal Plain deposits, are deep and well drained.

Area of Tidal Marshes.

51

Jersey, so has it determined the present landforms of the Inner and Outer Coastal Plains. In addition, changes in sea level between glacial and interglacial periods, combined with continuing wave erosion and deposition, have sculptured the coastal area of New Jersey.

As a result of many natural processes, the present soils of the Coastal Plain are very varied in their mixtures of clays, silts, sands, and gravels. The difference among the four rests in the size of the particles—clay being the smallest size and gravel the largest of the four. The Inner Coastal Plain has a larger proportion of clay in its soil than does the Outer Coastal Plain, which is much sandier. Also, on the Inner Coastal Plain there are deposits of marl, which is commonly called greensand. It consists of a sand-size aggregate of the green-colored mineral glauconite.

Some of the Coastal Plain soils provide better growing conditions for plants than others. For the most part, the soils of the Inner Coastal Plain are more fertile than those of the Outer Coastal Plain, which contain a much higher amount of quartz sand. Although the Outer Coastal Plain does have some fertile soils, the largest part of it, nearly 2,000 square miles, is made up of very sandy soil of low fertility which retains little of the moisture needed for plant growth. This area is known as the Pine Barrens (Figure 2-4).

On both the Inner and Outer Coastal Plains, wide variation in drainage conditions can be observed. There are extensive areas of wetland caused by high water tables, and at the other extreme there are very dry, sandy soil conditions. Finally, on the coast from Sandy Hook south to Cape May, one finds a narrow strip of sandy beaches that are sometimes separated from the mainland by bays, tidal ponds, and marshland.

Summary

From Precambrian time to the present, varied and complex geologic processes have produced a landscape of startling diversity in New Jersey considering the small size of the state (Figures 2-4, 2-5 and 2-6). Within its borders, New Jersey exhibits many variations in relief, in soil parent rock and surface deposit materials, in soil types, and in land drainage; all these have an impact on vegetation (Figures 2-7 and 2-8).

REFERENCES AND SOURCE MATERIAL

Kummel, H. B. 1940. The Geology of New Jersey. New Jersey Department of Conservation and Development. Geological Series Bulletin 50.

Quakenbush, G. A. 1955. Our New Jersey Land. New Jersey Agricultural Experimental Station Bulletin 775. New Brunswick, N.J.

Schuberth, C. J. 1968. The Geology of New York City and Environs. Natural History Press, Garden City, N.Y.

Tedrow, J. C. F. 1963. New Jersey Soils. Rutgers College of Agriculture, New Brunswick, N.J. Circular 601.

Widmer, Kemble. 1964. The Geology and Geography of New Jersey. Van Nostrand Co., Princeton, N.J.

3

Climate of New Jersey

Introduction

Located at a latitude between the 39th and 41st parallels, New Jersey is in the same north-south position as areas of northern California, southern Italy, and Turkey, but climatic conditions in the four places are quite different. This is evidence that latitudinal position is only one of a number of factors that determine the climate of a region. In addition to the angle of the sun's rays and the length of daylight — factors determined by latitude — an area's altitude and nearness to oceans and mountain ranges have an influence on its climate. Also important are the source and direction of air masses that flow over the region. In this chapter the climatic conditions that affect vegetation in New Jersey are described; before doing this, the reasons for the particular climatic conditions are explained.

Why the Climate of New Jersey Is What It Is

One might reasonably expect that New Jersey, with a coastline of about 125 miles, would have the type of climate known as maritime in which there is little change between summer and winter temperatures and little variation in daily temperatures. Areas in California, for example, at the same latitude as New Jersey and adjacent to the Pacific, have a relatively uniform climate throughout the year and only small variations in daily temperature. But, in contrast, New Jersey has a continental type of climate, one more typical of midwestern states.

A continental climate is characterized by a significant variation between the temperatures of summer and winter and by relatively large daily and day-to-day temperature fluctuations. In all parts of New Jersey there is a difference of more than 40°F in temperature from the warmest to the coldest month of the year, reflecting the continental nature of the climate. The reason for this is the direction of the prevailing winds. In winters the winds are from the northwest and cold air masses from subpolar areas of Canada move over New Jersey (Figure 3-1). In the months from May through September, again because of the prevailing winds, the state is blanketed with moist tropical air masses which originate over the Gulf of Mexico, flow inland, and then travel over very warm land before reaching New Jersey. California weather, on the other hand, is more uniform because the air-mass exchange is from ocean to land daily and from land to ocean nightly. The same air-mass exchange also occurs on a seasonal cycle and causes a moderating effect on the air temperature of the land, making it cooler in the summer (and in the daytime) and warmer in winter (and at night) than otherwise would be expected.

Temperatures of New Jersey

The annual temperature in New Jersey averages about 52°F for a normal year but, from the standpoint of vegetation, two considerations are important—the variations by locale within the state and the deviations from the average annual temperature from month to month.

For reporting of its weather data the United States government groups the official weather reporting stations in New Jersey in three classes—Northern, Southern, and Coastal stations. All weather stations slightly north of the line that separates the Piedmont from the Inner Coastal Plain are classed as Northern and with only four exceptions, all those south of the line are grouped as the Southern stations; the exceptions include the

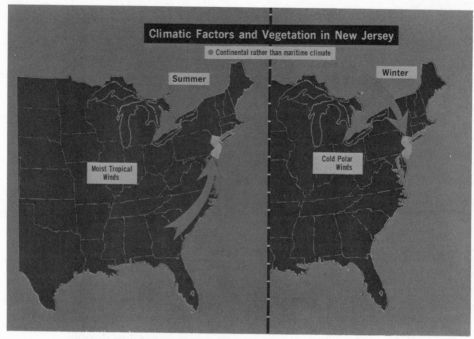

Figure 3-1 The direction of air mass movement over New Jersey accounts for the continental nature of its climate. Redrawn from Erwin Biel. The Climate of New Jersey. *In* The Economy of New Jersey. Rutgers University Press, New Brunswick, New Jersey, 1958.

weather stations at Cape May, Atlantic City, Sandy Hook, and Long Branch, which make up the third category, the Coastal group. The monthly average temperatures for each of the three groups are plotted in Figure 3-2. Two characteristics of New Jersey temperatures are apparent from these figures. First, the statistics reflect the continental nature of the climate by the wide variation between January, normally the coldest month, and July, the warmest month. Second, it is evident that the temperature differences between northern and southern locales in the state are greater in winter than in summer.

In January, the temperature of the Northern New Jersey stations averages 30.2°F, of the Southern 33.4°F, and of the Coastal 34.5°F. In July, the temperature averages 74.0°F in the

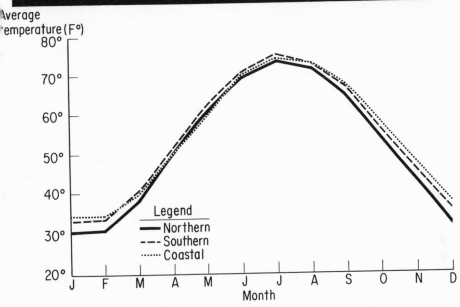

Figure 3-2 Monthly average temperatures recorded by weather stations in the northern, southern, and coastal sections of New Jersey. Source of data: Climatological Data, New Jersey. 1971 Annual Summary. Vol. 76, No. 13. U.S. Dept. of Commerce. Environmental Sciences Services Administration.

north, 75.4°F in the south, and 74.6°F on the coast. The Coastal stations in both cases reflect the moderating influence of the ocean. At extreme ends of the state the differences in winter temperatures are even more pronounced with January averaging as low as 26°F in the far northwest and over 34°F in the southeastern part of the state.

The winter variations between the northern and southern sections of New Jersey are particularly important from the standpoint of vegetation when one considers differences in the average length of the vegetative growing season—that is, the number of consecutive days on which the average temperature is about 43°F,

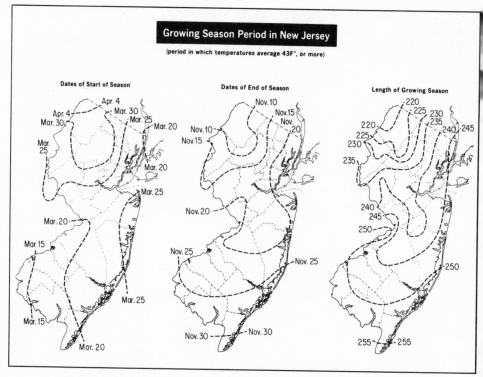

Figure 3-3   Dates of the start and the end of the growing season in New Jersey and the length in days of this season. The growing season is the period in which temperatures average 43°F, or more. Map redrawn from Erwin Biel. The Climate of New Jersey. *In* The Economy of New Jersey. Rutgers University Press, New Brunswick, New Jersey, 1958.

the point above which it is believed that most plant growth starts. Erwin Biel, in a description of the climate of New Jersey, describes the differences that occur in the growing season in New Jersey in terms of the starting date, the ending date, and the length of the period in which the daily temperature averages 43°F or more. Figure 3-3 plots the information in graphic form on a New Jersey county map. As the data indicate, the temperatures in the southern part of New Jersey permit plant growth to start earlier and to continue later in the fall, making the duration of the growing season as much as five weeks longer in the south.

Another way of looking at the temperature differences be-

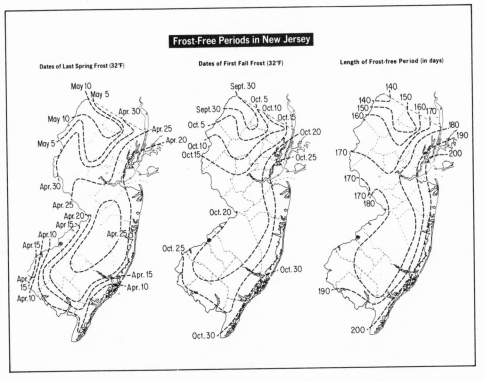

Figure 3-4   Dates of the last spring frost (32°F) and first fall frost (32°F) and the length of the frost-free period in days in New Jersey. Map redrawn from Climate and Man. 1941 Yearbook of Agriculture. U.S. Department of Agriculture.

tween northern and southern New Jersey is in terms of the length of the frostfree period, the interval in which the daily temperatures remain above 32°F. Maps showing the average dates of the last killing frost in the spring and the first killing frost in the fall and the duration of the intervening period without killing frost have been prepared by the U.S. Department of Agriculture; copies are shown in Figure 3-4. The length of the frostfree period in northernmost New Jersey is only 140 days as compared with 200 days in the Cape May area of southern New Jersey. Thus, there is much less chance in South Jersey that a frost-kill day will occur during the period of vegetation growth.

Both sets of statistics point up the "biological" differences between the climate of North and South Jersey. One comparison made is that Cape May in southern New Jersey has a normal January temperature about equal to that of southern Virginia; in contrast, the normal January temperatures in the most northern extremity of New Jersey are about equal to those in northern Ohio. The impact of these differences in the cold weather temperatures on vegetation is discussed in Part III.

Several other variations in temperature occur within the state. Ecologists have observed significant climatic differences on slopes of different exposures. For example, differences in temperature between the north-facing and south-facing slopes of Cushetunk Mountain, 600 feet high in central New Jersey, are enough to cause differences in the natural vegetation growing on the two slopes.

Variations in climate also occur between cities and surrounding countrysides as reflected in differences in temperature, humidity, precipitation, fog, and wind speed as well as in atmospheric pollution. That the temperatures in cities tend to be higher than in surrounding areas is explained by the combined result of city activities and substitution of building structures and pavement for vegetation. Urban concentration of people means a concentration of heat-creating sources whether required for the operation of heating or air conditioning systems or of industrial plants and vehicles. This heat as well as solar radiation is reflected and stored by pavements and building structures. Weather Bureau data indicate that the highest average annual temperature recorded by the weather stations in the northern New Jersey area occurs in Newark. Here the temperatures are particularly higher in the months of July and August; in the summer months the average temperatures in Newark are almost as high as in areas much further south in New Jersey.

In addition to temperature differences, there are other climatic variations between cities and their less settled surroundings. The former director of climatology of the U.S. Weather Bureau, Helmut Landsberg, estimates that cities as compared with surrounding countryside have 10 percent more precipitation, 10 percent more cloudiness, 30 percent more fog in summer, and 100 percent more fog in winter.

Precipitation in New Jersey

As compared with most other regions in the United States and in the world, New Jersey has a relatively high rainfall. The annual precipitation ranges from an average low of about 40 inches at the southeastern corner of the state to over 48 inches in two areas in the state, one in the north and the other in the south (Figure 3-5). Both places seem to receive the heaviest brunt of storms coming up the coast from tropical areas. Throughout the state rainfall normally is distributed pretty uniformly over the twelve months with slightly more falling in the months of July and August and less in February. Also important for vegetation is the fact that over a long period of time there have been relatively slight fluctuations in the total amount of rain recorded from year to year.

The official precipitation figures for the northern, southern and coastal sections of the state as defined are plotted in Figure 3-6. Northern New Jersey normally averages 46.96 inches of precipitation in a year, southern New Jersey 44.95 inches, and the coastal area 42.94 inches. Most of the difference occurs in the growing season, April through July. However, differences that do exist in the amount of rainfall between sections of New Jersey appear to be less important to vegetation than the water-holding capacity of different soil types and the varying water table levels. The effect of these on vegetation is discussed in Chapter 6.

In winter nearly 50 inches of snow normally falls in northern New Jersey while in the south there is an average of only about 14 inches. New Jersey is located in a belt with a relatively high frequency of sleet and glaze storms; these have a damaging influence on vegetation, particularly in forests where limbs or whole trees can be destroyed by the weight of ice.

Light

The length of daylight in New Jersey as measured at New Brunswick varies from about 9½ hours in January to nearly 12 hours in March and in September, and the longest day of the year, 15 hours, occurs in June. Of the total daylight hours, it is esti-

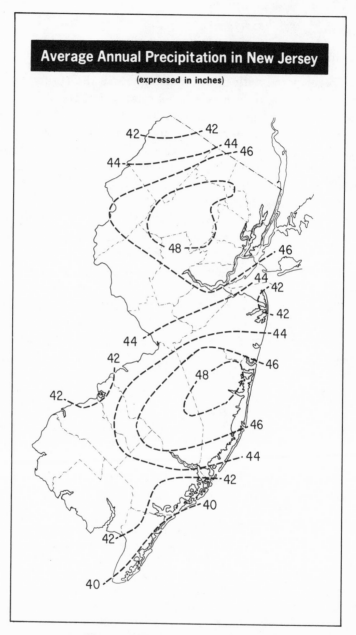

Figure 3-5   Total annual precipitation varying from 40″ to 48″ plotted by distribution in New Jersey. Map redrawn from Climate and Man. 1941 Yearbook of Agriculture. U.S. Department of Agriculture.

62

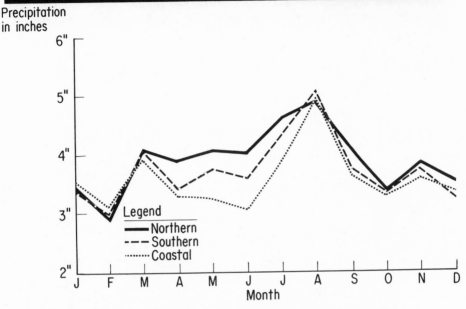

Figure 3-6 Monthly average precipitation (in inches) recorded by weather stations in northern, southern, and coastal sections of New Jersey. Source of data: Climatological Data, New Jersey. 1971 Annual Summary. Vol. 76, No. 13. U.S. Dept. of Commerce. Environmental Science Services Administration.

mated that in the nonpolluted areas of the state, the percentage of clear sunny days to total days averages 65 percent in the summer, 60 percent in the autumn and spring, and 55 percent in the winter. Before concentrated industrial areas were developed, there was not much variation within the state in the amount of sunshine. Now skies over industrial areas frequently become clouded with pollution wastes when less highly developed locales in the state are enjoying sunshine. The impact of atmospheric pollution on vegetation is discussed in the next chapter which deals with man's impact on vegetation in New Jersey.

Summary

Within the State of New Jersey there are variations in climatic conditions of temperature, precipitation, and light. Of these, the most important from the standpoint of vegetation is the variation in temperature as it influences the length of the growing season and the duration of the frostfree period.

REFERENCES AND SOURCE MATERIAL

Biel, Erwin R. 1958. The Climate of New Jersey. *In* The Economy of New Jersey. Rutgers University Press, New Brunswick, N.J. Pp. 53–98.

Cantlon, J. E. 1953. Vegetation and Microclimates on North and South Slopes of Cushetunk Mountain, New Jersey. Ecological Monographs 23: 241–270.

Lowry, William. 1967. The Climate of Cities. Scientific American 217: 2. Pp. 15–23.

U.S. Department of Agriculture. 1941. Climate and Man; 1941 Yearbook of Agriculture. Pp. 1002–1010.

U.S. Department of Commerce. 1971. Environmental Science Services Administration. Climatological Data, New Jersey; Annual Summary. Vol. 76, No. 13.

4

Man's Impact on the
Vegetation of New Jersey

Introduction

To describe the urbanized area that now stretches from southern New Hampshire southward through New Jersey to northern Virginia, Jean Gottmann suggested the word "megalopolis," which means very large city. In his book entitled *Megalopolis,* Gottmann states:

Nature's inheritance has been used constantly by the people in Megalopolis, though at each time and place the use has fitted into a system of relationships that have themselves been constantly varying.

The purpose of this chapter is to pinpoint the specific ways in which man has used, and misused, the landscape of New Jersey and the resulting impact on the natural vegetation. To do this, it is necessary that we look back into history and examine the relationships between man and vegetation that have existed since the day that man first arrived in the state.

Vegetation of New Jersey at the Time
of European Discovery

The date of the first arrival of man or his ancestors in New Jersey has not yet been firmly established. While his predecessors were probably here hundreds of thousands of years ago, man as we know him today is thought to have arrived in the eastern United States only as early as 10,000 or 11,000 years ago — a few thousand years after the last glacial ice had disappeared from New Jersey. The climate then is thought to have been a good deal cooler than at present, and although dinosaurs no longer roamed the state, the now extinct mastodons, close relatives to the elephants, were plentiful.

The first real evidence of man in the state is derived from Indian artifacts found along the upper Delaware River valley. These have been scientifically dated, indicating their use about 7,000 years ago. There is no accurate record of the number of Indians who lived in New Jersey before the Europeans came into the state; some authorities believe that there was a permanent Indian population of only 2,000 or 3,000 but others think that it was as high as 7,000 to 10,000. The Indians who occupied New Jersey at the time of the European settlement belonged to the tribe known as the Lenapes (or Delawares). Their numbers were distributed unevenly through the state with the greatest concentration in sites accessible by water, such as the valleys of the Delaware, Passaic, Hackensack, and Raritan rivers, the same areas that later were to be crowded by the European settlers.

In the course of living in New Jersey, the Indians, like their predecessor, prehistoric man, disturbed the natural vegetation. Sites had to be cleared for villages and for the cultivation of agricultural crops, particularly maize. According to Gordon Day, who studied the impact of Indian activities on the forests of northeastern United States, extensive cutting of forests was done by the Indians to provide the wood and bark used for utensils, weapons, canoes, shelters, and especially for fuel.

In addition to destroying the forest in many locales, the Indians modified its composition in much larger areas, mostly by setting fires. They recognized that many types of game such as deer prefer open woods to dense forests. Also, hunting can be made easier by the use of fire to drive game. Travel to and from hunting and fishing areas throughout the state was made easier when dense thickets of forests were cleared by burning. Once fires were started, they were allowed to burn until extinguished by physical barriers or by other natural means.

The Lenape Indians deliberately burned the woods in spring and fall and many early explorers of the New World including Henry Hudson observed smoking forests when approaching the coast. The report of a Dutch navigator nearing the coast of New Jersey in 1632 stated that those on shipboard

smelt the land, which gave off a sweet perfume, as the wind was from the Northwest, which blew off the land, and caused these sweet odors. This comes from the Indians setting fire, at this time of year, to the woods and thickets, in order to hunt; and the land is full of sweet smelling herbs, as sassafras, which has a sweet smell. When the wind blows out of the northwest, and the smoke is driven out to sea, it happens that the land is smelt before it is seen. The land can be seen when in from thirteen to fourteen fathoms.

This is one of several similar observations collected in a book of narratives of early Pennsylvania, west New Jersey, and Delaware by Albert Myers.

The results of the Indian-set fires in New Jersey are believed to be similar to those in New England which were described by Thomas Morton in 1637:

The Salvages are accustomed, to set fire of the Country in all places where they come; and to burne it, twize a year, vixe at the Springe, and the fall of the leafe. The reason that mooves them to doe so, is because it would other wise be so overgrowne with underweedes, that it would be all a copice wood, and the people would not be able in any wise to passe through the Country out of a beaten path . . . for this custome hath bin continued from the beginning. . . . For when the fire is once kindled, it dilates and spreads it selfe as well against, as with the winde; burning continually night and day, until a shower of raine falls to quench it. And this custome of firing the Country is the meanes to make it passable, and by the meanes the trees growe here, and there as in our parks.

Thus, contrary to popular opinion, the first European settlers to come to New Jersey or to other states on the East Coast did not find a vast expanse of virgin forest. This was particularly true of the accessible river valley areas that had been settled by the Indians. A good part of the land not cleared for villages or for the growing of crops was parklike with considerable openings in the wooded areas. Because some tree species are more resistant to fire damage than others, the effect of the Indian-set fires was to modify the composition of forests.

European Colonization and Its Impact on Vegetation

The Indians and their indiscriminate use of fire had a marked impact on the natural vegetation of New Jersey; it was the European settlers and their descendants, however, who truly disturbed the vegetation of the state.

The Delaware River valley was the first area settled by the Europeans for they, like the Indians, favored sites accessible to water. New Jersey, located as it is between two large rivers, the Hudson on the east and the Delaware on the west, offered relatively easy access and invited early settlement. As early as 1620 the Dutch started settling along the shores of the Hudson River and the Swedes along the Delaware, but the land of the state was officially claimed by England in 1664. Thereafter settlement of the state rapidly increased.

Colonization proceeded at different rates. Pioneers spread out from the initial sites of settlement, the Delaware, Hudson, Hackensack, Passaic, and Raritan River valleys, and occupied the rest of the lowlands in the Piedmont and Inner Coastal Plain sections (Figure 2-3). Although it was both accessible by water and fertile, the Kittatinny Valley was not settled initially because it was occupied by hostile Iroquois Indians.

Very soon most of the upland on the Piedmont and Inner Coastal Plain was cleared for agricultural use. The forests on higher slopes or ridges or in areas too wet for cultivation were kept as the source of wood needed for shelter, fencing, household utensils, and heating. From the time of initial settlement, however, domestic animals such as cattle, horses, and hogs were allowed to graze freely in the woodlands.

By 1726, the date of the earliest census, the population of New Jersey was 32,442 and by 1784 this had jumped to 149,435. To facilitate clearing of land for settlement or for cultivation, the Indian practice of setting fire to the forests was continued by some European settlers. By the time New Jersey became a state in 1778, no extensive areas of land well suited to farming remained wooded in the central part of the state.

Nineteenth Century: Man and Vegetation in New Jersey

As the population of New Jersey continued to grow, so did the need for wood. Until the middle of the nineteenth century, wood was the only source of fuel. In addition to accommodating the household needs of the growing population, wood was needed in enormous quantities as fuel for the operation of steamboats and locomotives, and for the early New Jersey industries of lumbering, iron making, charcoal production, and the manufacture of glass. In 1783 a visitor from Europe observed that a single iron furnace in Union, New Jersey, had exhausted a forest of nearly 20,000 acres in about twelve or fifteen years, and the works had to be abandoned for lack of wood.

Settlers spread out into less fertile areas of the Highlands and the Ridge and Valley sections but because of the poor soil conditions, agricultural subsistence on the ridges never was really successful. For the same reason the major part of the Outer Coastal Plain, the area of the Pine Barrens (Figure 2-4), was not attractive to the early settlers. However, to meet the enormous demand for wood, the forests in these areas were cut frequently and repeatedly for cordwood. The indiscriminate cutting of woodlands at twenty- to twenty-five-year intervals was accompanied by continuing damage by wild fires. In 1874 an estimated 100,000 acres of forests were burned and in 1885 another 128,000 acres.

Fortunately, with the introduction of coal in 1850 and other fuel material later, the demand for wood from the forests of New Jersey decreased. However, repeated cutting of wooded areas

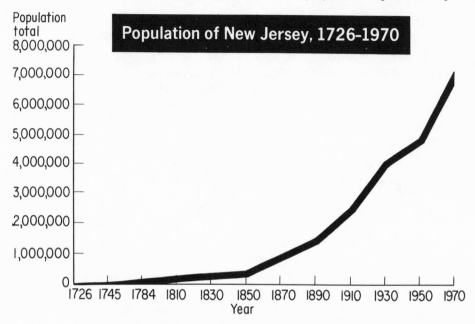

Figure 4-1 Growth of population in New Jersey from 1726 to 1970. Source of data: Population figures for 1726–1950 from John Brush. The Population of New Jersey. Rutgers University Press, New Brunswick, New Jersey, 1956. Figures for later years from the New Jersey Department of Labor and Industry, Bureau of Research and Statistics.

combined with fire damage had already so altered the composition of the forests that the effects can be seen in New Jersey today.

Starting in 1850, the population of New Jersey increased rapidly and grew from a total of 489,000 in 1850 to slightly under 2 million by 1900. During this time, however, because of the substitution of anthracite coal for wood and charcoal, man did less damage to the natural vegetation of New Jersey than in the preceding fifty years. That the forests of the state were able to recover, in part, from some of the past misuses is reflected in the 1899 report of the New Jersey State Geologist:

Reviewing all the evidence which we have collected, therefore, we state with confidence that there was progressive cutting and clearing-up of the original forest all over the State, from its settlement until 1860, but that at the latter date very little original forest remained. This cutting was most severe about 1850, and from 1850 to 1860 was the period of maximum deforestation. The forest was then younger and smaller, with a larger proportion of stump and brush land than has prevailed at any time since. The cutting of very young growth has decreased to a marked extent in recent years, and the average size and age of the forest has increased. At present not two per cent of the forest is cut annually, so that at the present rate of cutting all of the forest may attain an age of fifty years.

In the same report the State Geologist stated that in 1899 46 percent of New Jersey land was covered by forest—as much as had existed in 1860. An almost equal amount of land was used for farming.

As might be expected, even in 1899 the wooded areas were not distributed evenly throughout the state. At this time, only 15 percent of the Inner Coastal Plain and 23 percent of the Piedmont area consisted of wooded land. In contrast, 68 percent of the Outer Coastal Plain, 56 percent of the Highlands, and 45 percent of the Ridge and Valley section were forested.

Twentieth Century: Man and Vegetation in New Jersey

From 1900 to 1970 the population of New Jersey more than tripled (Figure 4-1), and New Jersey has become known as the most urban state of the union. The statistics that support this reputation are staggering: New Jersey has a population density averaging 954 people per square mile, the highest of the fifty states; 8.9 percent of the total population of the United States lives within a 50-mile radius of New Brunswick, located in central New Jersey; and of the twelve most densely populated urban places of 25,000 or more population in the United States, eight are located in New Jersey. Fortunately for vegetation as well as for man himself, the past increases in population have tended to follow the distribution pattern established originally by the Indians and followed by European colonists. Certain areas of the state have been heavily settled while others remain sparsely populated. Two sets of statistics highlight this fact: the popu-

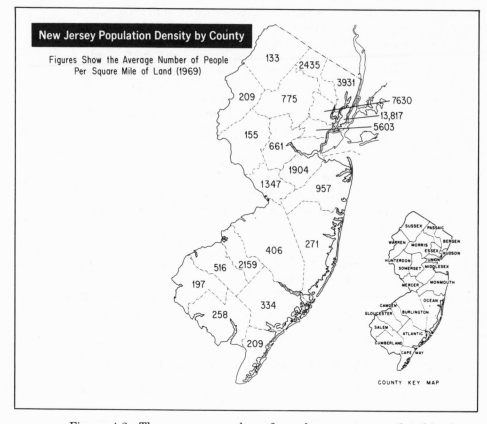

Figure 4-2 The average number of people per square mile of land (population density) in each county of New Jersey in 1969. Source of data: 1970 Population Estimates. New Jersey Department of Labor and Industry, Bureau of Research and Statistics.

lation densities by county, shown in Figure 4-2; and the county percentages of total population as compared with the percentages of total land area, shown in Figure 4-3. Both sets of figures indicate that as of 1970 the population of New Jersey is concentrated in three areas.

1. By far the highest concentration of people is in the six counties that border on the New York metropolitan region: Bergen, Passaic, Hudson, Essex, Union, and Middlesex. In 1970, the population of the six counties totaled 3,980,730, or 56.2 percent

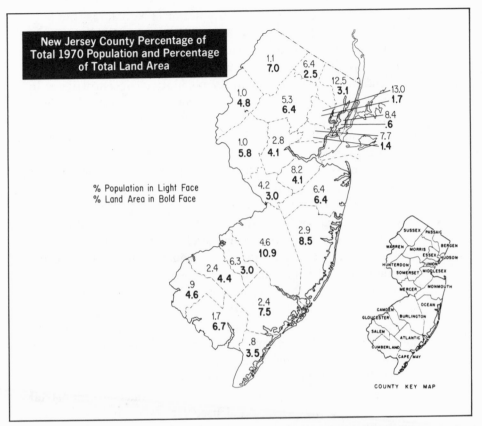

Figure 4-3   A comparison for each New Jersey county of its percentages of the total population of the state and of the total land area of the state. Source of data: County Data Sheets, 1970. New Jersey Department of Labor and Industry.

of the total population of the state, although these counties account for only 13.4 percent of the total land in the state.

2. A much less concentrated center is in Camden, the county that adjoins the Philadelphia area. Camden has 6.3 percent of the population but only 3 percent of the land area of the state.

3. Population is also concentrated though to a still lesser extent in the Trenton, Mercer County area, which has 4.2 percent of the population but only 3 percent of the land.

Altogether, the eight counties enumerated above account

for about 67 percent of the total population of New Jersey but only about 19 percent of the land area. Of the remaiming thirteen counties in New Jersey, all but four have population densities of less than 500 people per square mile.

Another way of looking at the population concentration is in terms of the five physiographic sections of the state. Dr. John Brush of Rutgers University developed such figures based on the distribution of the population in 1950. The results of his studies are plotted in Figure 2-1.

It is because the population is not equally distributed that natural areas remain in New Jersey in spite of its high degree of urbanization. How long this will remain true is questionable; since 1960 areas of the state that were sparsely populated have begun to grow, some at a very accelerated rate. For example, Ocean County won what might be considered an unfortunate distinction, that of being the fastest growing county in the whole United States in the decade 1960–1970. In this same decade other counties that historically have also been sparsely populated — Burlington, Sussex, and Morris — grew at a much more accelerated rate than did the more urbanized counties of Bergen, Passaic, Essex, Hudson, Union, and Mercer.

To accommodate the increasing population in New Jersey, more and more land is being devoted to residences, industrial buildings, and highways. Land used for these purposes is generally stripped completely of its natural vegetation. In view of this, it is striking that according to state forestry reports in 1970 the total forest area of New Jersey is about equal to that of 1860 and of 1900, when it was estimated that 46 percent of the land of New Jersey was wooded. That so large a percentage is wooded is surprising since the population in 1970 was more than eight times that of 1860 and four times that of 1890.

There are several reasons for the continued high percentage of forest land despite explosive population growth. First, already mentioned, the population increase has been concentrated in the northeast counties adjacent to metropolitan New York. Even in 1899 these counties did not have a large percentage of their land area in forest cover (Figure 4-4). Second, there has been a sizable decrease in the amount of land for farming. In 1899, 46 percent of the land was devoted to farming activities, but according to most recent statistics, only 24 percent of New Jersey remains as farmland. As described in Chapter 1, land once cultivated and then abandoned reverts to woodland naturally. This process has helped to offset the appropriation of other forest land for development. Third, the amount of forest land conserved as part of state-owned parks and recreation areas has increased in the last fifty years.

In 1899 there were six counties in New Jersey that were more than 50 percent forested. Five of these were the southeastern counties of Cumberland, Cape May, Atlantic, Burlington, and Ocean, and the sixth was Passaic in the northeast (Figure 4-4). By 1970 five of the six counties (all but Cumberland) remained more than 50 percent forested and the amount of woodland in two other counties — Sussex and Morris — had increased so that more than 50 percent of their land was also in forests. For the most part, the counties that were sparsely wooded in 1899 remain so to this day but in six counties in addition to Sussex and Morris counties — Warren, Hunterdon, Mercer, Monmouth, Somerset and Salem — the percentage of land in forest is greater in 1970 than it was seventy years previously.

On the whole, however, the locations of forest areas of New Jersey are concentrated much as they were as early as 1800. Obviously, then as now, the concentration of woodlands is in areas of the state not highly settled. At present there are about 2,120,000 acres of forest land in New Jersey; the distribution of this total by county is shown in Figure 4-5. As indicated by the figures, the largest concentration of forest land in the state, 47 percent of the total, is in the five southeastern counties of Burlington, Ocean, Atlantic, Cape May, and Cumberland. This is the area of the Pine Barrens.

The second concentrated area of forest land, smaller than the Pine Barrens, is in the northern part of the state — the region of higher elevations in New Jersey including the ridges of the northern Highlands and Kittatinny Mountain in the Ridge and Valley section. In the Piedmont section, generally the only areas that

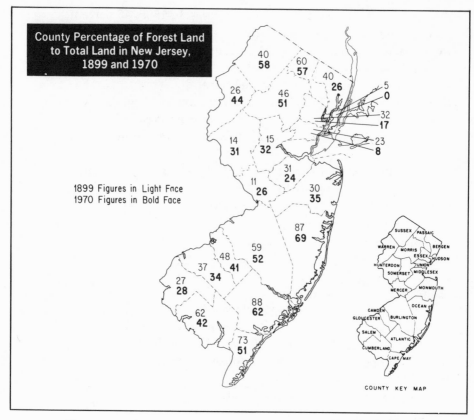

Figure 4-4   A comparison for each New Jersey county of the percentage of its total land in forest for the two years, 1899 and 1970. Source of data: Figures for 1899 taken from C. C. Vermeule. Report on Forests in Annual Report of the State Geologist for 1899, Trenton, New Jersey. Figures for 1970 supplied by the Northeastern Forest Experimental Station, Upper Darby, Pennsylvania.

are still wooded are diabase and basaltic ridges, swamp lowland areas, and state parks. On the Inner Coastal Plain, because of urbanization and use of the fertile soils for agriculture, only the more poorly drained areas remain wooded. The distribution of forest land just described is illustrated graphically by a map prepared in by the office of the New Jersey State Forester (Figure 4-6).

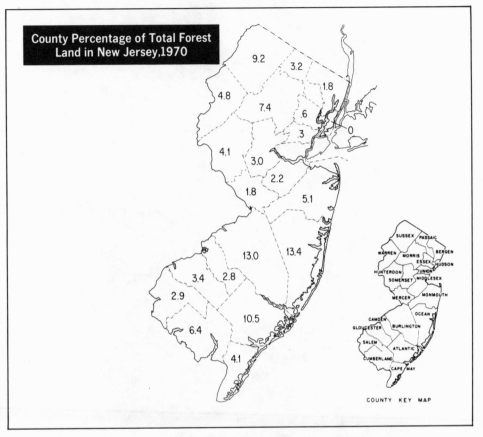

Figure 4-5   The percentage of the total state forest land that each New Jersey county had in 1970. Source of data: Northeastern Forest Experimental Station, Upper Darby, Pennsylvania.

Thus far, attention has been paid only to that part of New Jersey land which potentially would have forest growth as its natural vegetation if the land were left undisturbed by man. This includes all but a small part of the state. Of the 7,509 square miles of land in New Jersey now, only 353 square miles, or 4.5 percent of the state, consist of tidal land, or salt marshland on which a nonforest type of natural vegetation prevails. The degree to which man has invaded and destroyed much of the orig-

Figure 4-6 A forest cover map of New Jersey showing forest areas in black. Map was prepared by the Office of the New Jersey State Forester, New Jersey Department of Environmental Protection in 1969; the source data for the map was U.S. topographic maps prepared in earlier years.

inal marshland can be observed when riding through the Newark and Jersey City area. By drainage systems or by landfill operations much of the natural marshland in the region has been destroyed and that which remains is threatened by pollution.

Vegetation and Pollution

It was the Industrial Revolution that really caused the start of excessive pollution of air and water. It is said that although air pollution damage to vegetation in New Jersey dates back about one hundred years, the problem did not become acute until the late 1940s. Since then, continued industrial expansion and the increased use of automobiles have led to higher and higher pollution levels in the state. Now, Rutgers plant pathologists report air-pollution damage to more than sixty types of plants growing in every county of the state.

Certain types and levels of pollution cause serious medical problems to human beings, but even levels of pollution seemingly not harmful to man have caused serious injury to plants. For this reason, it is thought that, generally, plants have lower tolerances to pollutants, particularly those in the atmosphere, than do people.

Plants filter out dust, soot, and fly ash from the air. But too much pollutant material absorbed from the atmosphere or from the soil through root hairs can be toxic to a plant. Plant species differ in tolerance of pollutants. Some species of trees, shrubs, and even agricultural crops show no sensitivity to certain types or doses of pollutants that will cause serious injury or even death to other types of plants. Damage to plants in New Jersey stems principally from acid gases in the atmosphere (sulfur dioxide, chlorine, and hydrogen fluoride) and from the so-called oxidants that come primarily from automobile exhausts, the latter including ozone, nitrogen oxide, and a toxic mixture called PAN. Rutgers plant scientists report that in New Jersey at this time a heavy accumulation of ozone in the atmosphere is causing more damage to plants than any other atmospheric pollutant. It has threatened not only natural plant growth but also commercial production of several agricultural crops such as spinach, endive, and chickory.

The natural vegetation in New Jersey has suffered from other types of pollution. Scientists have found that atmospheric emissions of zinc and cadmium from industrial plants accumulate in the surrounding soils and affect plants. Also, toxic wastes deposited in stream waters from industrial activities have reduced the variety of plant species growing in streams and along stream banks. Finally, the impact of the deposits of residential sewer wastes in stream waters of the state as well as the effect of excessive fertilizer runoffs from upland soils can be seen all over the state. The green scum on polluted waters actually is an aggregate of small living plants called algae. Algae have abnormal growth in situations of excessive nutrients in the water; they then choke out other forms of plant life and fish.

Summary

Since his initial occupation of New Jersey, man has proceeded to destroy totally or to modify the natural vegetation of the state. The forest primeval has long been gone. More than half of the state previously covered by forests or marshes is now paved, settled with houses or industrial buildings, or devoted to farming activities. In the past the population has been highly concentrated in the most accessible and, fortunately, least diverse land of the state. This explains why New Jersey, the most densely populated state, continues to have diverse natural vegetation.

While most of the Piedmont and Inner Coastal Plain sections are highly developed and while the fertile valleys in the northern sections are used for farming, a large proportion of the land remains in forest. The woodlands are concentrated in the Pine Barrens on the Outer Coastal Plain and on the northern ridges. The present composition of these forests reflects, however, the impact of man's previous activities – particularly the results of fire damage and indiscriminate cutting. In later chapters the vegetation of New Jersey is described – what it is now and what it might have been without the interference of man.

REFERENCES AND SOURCE MATERIAL

Brush, John E. 1956. The Population of New Jersey. Rutgers University Press, New Brunswick, N.J.

Daines, R. H., I. A. Leone, and E. Brennan, 1960. Air Pollution as It Affects Agriculture in New Jersey. New Jersey Agricultural Experiment Station Bulletin 794. New Brunswick, N.J.

Day, Gordon M. 1953. The Indian as an Ecological Factor in the Northeastern Forest. Ecology 34: 329–346.

Gottmann, Jean. 1961. Megalopolis. MIT Press, Cambridge, Mass.

Moore, E. B. 1939. Forest Management in New Jersey. New Jersey Department of Conservation and Development, Trenton, N.J.

Morton, T. 1637. New English Canann. John Wilson and Son, 1883. Boston.

Muntz, Alfred P. 1959. The Changing Geography of the New Jersey Woodlands. Ph.D. Thesis, University of Wisconsin, Madison, Wisconsin.

Myers, Albert. 1912. Narratives of Early Pennsylvania, West New Jersey, and Delaware. Charles Scribner's Sons, New York.

Vermeule, C. C., A. Hollick, J. B. Smith, and G. Pinchot. 1900. Report on Forests in Annual Report of the State Geologist for 1899, Trenton, N.J.

5

Other Biological Influences: Plant-Plant and Animal-Plant Relationships in New Jersey

Introduction

The final component to be considered in the ecosystems of New Jersey is biological interrelationships other than those involving man — specifically, interrelationships among plants themselves and between plants and animals. In this chapter both past and present interrelationships that exercise a major influence on natural vegetation are described.

Past History of Vegetation; Plant-Plant Relationships

The earliest real evidence of plants in new Jersey is found in fossils, the remains of plants that lived a long time ago, and have been preserved in rock or swamp material. The time at which a fossilized plant lived is identified by scientific dating of the rock or other material in which the plant is embedded.

The earliest known forms of life existed in Precambrian time, more than 500 million years ago, yet few fossils have been found in the world dating back to this period. But Kemble Widmer, in a book on the geology of New Jersey, suggests that

82

unproved as it is in New Jersey, it would seem reasonable to assume from evidence elsewhere in the world that the graphite in the Franklin formation (rocks in the Highlands) is all that is left of very ancient Precambrian marine life which once existed in New Jersey.

Fossils of primitive plants and animals dating back to the Paleozoic era, more than 180 million years ago, have actually been found in New Jersey, though not as plentifully as in neighboring states. More fossils are available from Triassic times, the first period of the Mesozoic era, in which the climate for the most part was warm and moist. The type of flowering plants that we know today did not exist at that time. Instead, more primitive plants, some huge trees fernlike in appearance, grew in the swamps through which dinosaurs roamed, leaving their trace in the form of footprints embedded in the Triassic shale. It was in the later Cretaceous period, 63 to 125 million years ago, that plants somewhat similar but still ancestral to those of today grew in New Jersey. Fossils dating back to Tertiary times (1 million to 63 million years ago) give evidence that during this time, plants grew in New Jersey which were quite similar to those now found in the latitude of Virginia.

When the glacier ice started to move over the northern part of our state about 1 million years ago, it must have destroyed existing vegetation in a manner similar to that strikingly illustrated in a photograph taken recently in Alaska (Figure 5-1). Advancing ice completely overruns a forest, destroying all trees that lie in its path.

During the period of extensive glaciation in North America, some plant species were able to continue their existence by migrating southward. Others unable to migrate were completely exterminated.

While the boundaries of land covered by glacial ice in New Jersey have been clearly identified, there is not complete agreement about the climate or the vegetation that continued to exist south of the ice sheet. Some believe that it was so cold in southern New Jersey that only a treeless vegetation similar to that found in the arctic today could have existed. Others think that while either the arctic tundra type of vegetation or spruce-fir forests

Figure 5-1 Glacial ice sheet advancing over a forest. In 1966 the north margin of Meares glacier in south central Alaska overruns trees in a 300-year-old forest in the same manner that the ice sheets must have advanced on forests in New Jersey in Pleistocene time. *Photo by W. O. Field, American Geographical Society, taken September 2, 1966.*

now typical of Canada may have grown in a belt 10 to 50 miles wide adjacent to the glacial edge, a more temperate type of vegetation survived farther south in the state. In any event, there is enough evidence to conclude that because of the colder climate, vegetation different from that today must have existed in the part of New Jersey that was south of the ice sheets.

In the last 1 million years the glacial ice has alternately advanced and retreated depending upon climatic temperatures. In extended cold periods the glacial ice advances and in warmer periods, such as we are now enjoying, glacial ice melts, causing higher sea levels. As a result of the glacial ice melt, land areas previously covered with vegetation now are inundated by seawater. Evidence of the changing sea level has been found in the Hackensack Meadows where remains of a former cedar tree forest are now overlain with tidal marsh grass.

The last ice sheet, which at its maximum was as much as 2,500 to 3,500 feet in thickness, probably vanished from New Jersey between 10,000 and 15,000 years ago. After this date, vegetation gradually reestablished itself on those parts of the Ridge and Valley, Highlands, and Piedmont sections that had been covered by ice. In addition, the type of vegetation growing south of the ice sheet gradually changed its composition; most of the species now more typical of colder regions migrated northward and were replaced by those presently found in New Jersey. A record of changes in vegetation types of New Jersey is preserved in the peat bogs. As long as a bog has existed, pollen grains from the plants growing in and around the bog have been deposited in the bog peat which acts as a preservative. Thus, pollen from plants currently growing in the bog area accumulates in a layer above that from plants growing in previous times. By analyzing sample cores from the bog peats, scientists can reconstruct from the preserved pollen grains a history of the groups of plants that grew in the past. From the pollen-recorded sequence of changing vegetation types, the history of climatic changes that must have occurred in the past can be postulated. Some of the species now remaining as relics of the Ice Age and plant immigrants from areas farther south are identified in later chapters.

The long-term changes that occur in climate such as those just described are one reason for plant succession, the process (see Chapter 1) whereby whole groups of plants tend to displace others over a period of time. This happens because lower (or higher) temperatures are more favorable to some species than others. Thus, long-term cooling or warming trends have always been accompanied by gradual changes in the plant species composition of an area. The vegetation of New Jersey 10,000 years ago may have resembled that now present in the arctic or in Canada, but some 10,000 years in the future, if the warming trend continues, the present vegetation may be replaced by plant immigrants more typical of warmer climates such as those now in the southeastern United States.

Present Plant-Plant Relationships

On a much smaller time scale than the changes caused by long-term climatic fluctuations are successional changes in natural vegetation that result from some type of interference. In Chapter 1 an example of such a change was described — a piece of land was stripped of its natural vegetation, farmed, and then abandoned. The vegetation on the site gradually and naturally reverted back to forest land. For the most part it is competitive action among plants that causes this type of natural change in vegetation. The plants growing in any area usually are competing among themselves — individuals within a single species and whole populations of different species — for the available resources of water, soil nutrients, and light. Whenever one or more of these resources is in short supply, some plants, and even all individuals of a whole species, may not be as successful as other plants or other species in getting their share of the needed resource. Also, slight short-term changes in environmental conditions may shift the advantage of exploitation from one species to another. Competition for a particular resource may lead to the death of certain individuals or to the extinction of whole populations of certain species at the benefit of others. These failures and successes over a period of time result in the process of succession and cause gradual changes in the plant composition of a given area.

By his actions man may so disturb the environmental conditions of a site that he interrupts the natural sequence of successional development and starts a new series of successional changes. Such is the case of the example cited in Chapter 1. Successional stages of vegetation may originate from forces other than man; glaciation in New Jersey, for example, was responsible for the initiation of varied successional stages of vegetation because the ice sheets completely obliterated large areas of natural plant growth that probably represented the climax stage as defined in Chapter 1. Much of the land left bare of plant growth by glaciers thousands of years ago is now in various stages of successional growth. The changes in species composition as they specifically occur in succession in New Jersey are described in the fourth part of this book.

Destructive Plant Interrelationships

Some organisms classified as plants can cause disease and even death of other plants. One such plant organism is a particular type of fungus known as the chestnut blight. At the beginning of the twentieth century the American chestnut tree was a common and beautiful tree in the forests of central and northern New Jersey as well as other Eastern states. In 1904 the chestnut blight fungus was accidentally brought into the United States from eastern Asia. This fungus, a parasitic organism, that causes a disease in the tree bark can kill a chestnut tree. Within fifty years of its introduction into the country, the disease spread over the entire range of chestnut trees killing all mature trees growing in northeastern United States; today no fully grown chestnut tree remains in New Jersey forests. While sprouts may develop from diseased tree trunks, they rarely grow more than 15 to 20 feet high before being killed by the fungus. The die-off of the chestnut trees left gaping holes in the forests of the state, which are now being filled by other types of trees (Figure 5-2).

Today in New Jersey as well as in New England we can see

Light shines on the leaves of a chestnut tree sprout in High Point State Park. Tree sprout will grow only about 15 feet tall and then be killed back by fungus blight.

Dead elm trees on floodplain in Mercer County. The trees are killed by fungus carried from tree to tree by elm bark beetle.

Figure 5-2   Destructive plant interrelationships.

the disastrous results of another tree disease, the Dutch elm disease, caused by another type of fungus that is carried from tree to tree by the elm bark beetle. Not only are the huge elm trees that line our streets dying off in large numbers but those growing naturally in the woods are gradually disappearing (Figure 5-2).

Animal-Plant Relationships

In New Jersey there is abundant evidence of both beneficial and destructive interrelationships between plants and animals. An example of beneficial interrelationships is the dependence of forest trees upon animals for successful survival. Earthworms along with many other soil-inhabiting organisms improve the soil growing conditions for plants; still other organisms that are mostly microscopic decompose fallen logs and leaf litter in the forest, thereby returning to the soil nutrients needed for plant growth. Also, we can observe the role of flying insects, such as bees, butterflies, and moths, who aid in the perpetuation of native plants by carrying from plant to plant the pollen needed for reproduction. The result of successful reproduction are plant seeds which may be contained in berries or berrylike fruit such as that of the native dogwood trees or in nuts such as the acorns of the oak trees. The fruit-eating birds — the robin, blue jay, and red-eyed vireo, for example — aid the spread of dogwood and other trees throughout the woods of central and northern New Jersey. In the same way the common gray squirrel, who fails to eat all the acorns that he caches away, provides an opportunity for an acorn to develop into a tree in a location far beyond the range of its parent tree.

On the other hand, plant seeds or nuts may be destroyed as well as disseminated by animals. Scientists studying oak trees have found that each year a large proportion of the acorn crop is destroyed or badly damaged by insects. Many species of insects can be found in acorns but most of the damage is caused by several species of weevils, moths, wasps, and flies.

Other actions of animals have even more serious impact on the natural vegetation. Currently, the greatest damage being done to the natural vegetation of New Jersey stems from the explosive growth in populations of particular insects who feed on the foliage of our native trees. While some eating of plant parts including leaf foliage is always expected, complete defoliation of all trees of a species is abnormal and raises the tree mortality rate for the species far higher than that normally expected. In recent years there have been large increases in the population of three types of leaf-eating insects—the gypsy moth, the canker worm, and leaf roller. The gypsy moth particularly is currently having a devastating effect on the forests of New Jersey.

The gypsy moth was imported into Massachusetts from Europe in 1869 by a French scientist attempting to develop a strong race of silk-producing insects by crossing gypsy moths with silkworm moths. Unfortunately, in a windstorm gypsy moth eggs were accidentally blown out of the laboratory into the surrounding area and hatched into caterpillars which had no natural enemies to control their numbers. Since this catastrophe, the gypsy moth has spread into other Eastern states, including New Jersey.

The gypsy moth has four life stages—egg, larva (or caterpillar), pupa (resting stage), and adult (moth). It is the caterpillar life stage that causes severe damage to vegetation. Gypsy moth caterpillars feed heavily at night on tree foliage, in this state particularly on the oaks and the pines, and in a short time will completely defoliate a whole tree. Although a white oak tree can survive one complete defoliation, it becomes weakened and loses its natural growth. If defoliated in two successive years, the tree will usually die. The other common oaks in New Jersey—red, black, and scarlet—are slightly more resistant but also will die from successive defoliation. On the other hand, a single severe defoliation of a pine or hemlock tree may cause its death. Other species of trees in New Jersey also killed by gypsy moth defoliation include the beech, birch, willow, poplar, and red maple.

Although gypsy moth egg masses were discovered in New Jersey as early as 1919, only 5 acres of woodland were defoliated in 1966. By 1971 this figure had grown to 180,000 acres and whole forests are now being completely defoliated (Figure 5-3). An esti-

Figure 5-3 Example of a destructive animal-plant relationship: the defoliation of oak forest by the gypsy moth caterpillar on Sunrise Mountain, Sussex County, New Jersey, June 1969. *New Jersey Department of Conservation and Economic Development.*

mated 30 percent of the oak trees in the Morristown area alone have been lost and by the spring of 1971 over 1 million oak trees in the Newark watershed had been destroyed. Even more forest acres in the state were infested in 1972 but serious outbreaks of the gypsy moth caterpillar occurred in counties that heretofore had little problem with the insect damage.

In addition to damage by the gypsy moth, pine trees on the Outer Coastal Plain also suffer from the feeding of pine moth larvae which chew on their foliage. Altogether, it was estimated by the New Jersey Department of Forestry that in 1970 alone the mortality of forest trees caused by foliage-eating insects had increased from a normal of about 3 percent to as much as 40 percent in some locales. The oak trees particularly have been hard hit and it is feared that if the infestation of the leaf-eating insects continues at this high rate, there will be a serious impact on the natural forests of New Jersey.

Another leaf-eating insect is the so-called tent caterpillar which derives its name from the disagreeable looking large white webs made in the forks of branches of trees frequently growing along the roadsides of New Jersey. So far, these insects have been less damaging than the gypsy moth to natural forest areas only because they feed primarily on the foliage of wild cherry and other fruit trees which usually grow at the edge of mature forests.

Regular cycles of locust and cicada outbursts of population also are not uncommon in New Jersey. One such outbreak occurred in 1970 and in areas of the Piedmont caused some kill of small trees and loss of growth in older trees.

The growing evidence in New Jersey as in other states is that recently the more damaging of the insect population explosions are resulting from man's interference with the natural processes by which insect populations normally would be controlled. Usually after a few years of explosive growth, an animal population will naturally "crash" (or reduce in numbers) because of lack of food, attack by predators and parasites, climatic conditions, and disease. Ecologists believe that with the indiscriminate use of pesticides man interferes with these natural forces. For one thing, pesticides kill off predators and parasites that would feed on destructive insects. For this reason attempts are now being made to control by means other than chemical sprays the excessive outbreaks of gypsy moth and other damaging insects. In an area of the Jockey Hollow section of Morristown National Park, scientists from the New Jersey Department of Agriculture are successfully using imported enemies of the gypsy moth—a predaceous beetle species, parasitic wasps, and a toxic bacillus—

Figure 5-4   A beaver-created pond in Flatbrook Brook near High Point State Park. The beaver dam forms a dike (*at the left*) creating flooding of land which previously was dry enough for tree growth. Logs forming the beaver lodge can be seen in upper right corner of the pond.

to attack the egg, larval, and pupal stages of the moth. This is a form of biological control of insect populations. In addition, successful efforts have been made to reduce the effective mating of male and female gypsy moths by sterilizing males who are lured into traps by a synthetic female moth's sex attractant.

Beavers by building dams can change the drainage conditions along a stream; areas not previously flooded may be submerged by a beaver-created pond (Figure 5-4). This has an impact on the vegetation as described in Chapter 9. The only larger animal now causing extensive damage to our natural vegetation is deer. Without natural predators such as wolves to control its population, overpopulation of deer occurs even in New Jersey. In these years extensive damage can be done to forest vegetation in both northern and southern New Jersey by deer that nibble on shrubs, small trees, or branches of larger trees. Deer may even invade residential areas in search of food. However, through its wildlife management program, the state attempts to control the size of the white-tail deer herd which is now New Jersey's most important big game animal.

Summary

Interrelationships among plants as well as plant-animal interactions influence the natural vegetation of an area. In New Jersey long-term fluctuations in climatic conditions have resulted in changes in natural vegetation that reflect competitive successes and failures of particular species to particular environmental conditions. Also, man's actions and natural forces such as glaciation have caused changes in site conditions, thereby interrupting the natural development of vegetation. Continual competition among plants under altered environmental conditions results in successional stages of vegetation.

REFERENCES AND SOURCE MATERIAL

Buell, Murray F. 1970. Time of Origin of New Jersey Pine Barrens Bogs. Bulletin of the Torrey Botanical Club 97: 105–108.

Gibson, Lester P. 1972. Insects That Damage White Oak Acorns. U.S.D.A. Forest Service Research Paper NE-220. Northeastern Forest Experiment Station, Upper Darby, Pa.

Kegg, John D. 1970 and 1971. New Jersey Forest Pest Reporter, Vol. IV, Nos. 2 and 5. Results of 1971 Gypsy Moth Aerial Survey, N.J. Department of Agriculture. Division of Plant Industry, Trenton, N.J.

McCormick, Jack. 1966. The Life of the Forest. McGraw-Hill, New York.

Rogers, Georgia M. 1969 and 1970. The Menace of the Gypsy Moth *in* New Jersey Municipalities, May 1969 issue. The Gypsy Moth *in* New Jersey Municipalities, October 1970 issue.

Shafer, Elwood L., Jr. 1964. Deer Browsing of Hardwoods in the Northeast: A Review and Analysis of the Situation and the Research Needed. Northeast Forest Experiment Station, Upper Darby, Pa.

Vermeule, C. C., A. Hollick, J. B. Smith, and G. Pinchot. 1900. Report on Forests in Annual Report of the State Geologist for 1899, Trenton, N.J.

Widmer, Kemble. 1964. The Geology and Geography of New Jersey. Van Nostrand Co., Princeton, N.J.

Part III
Natural Landscape of New Jersey

6

Natural Landscape of New Jersey: Terrestrial Plant Habitats and Vegetation Types

Introduction

In Chapter 1 the ecosystem factors that influence vegetation of any region—geologic and soil features, climate, man's actions and other biological interrelationships—are identified, and reasons are given to explain why landscape variety or diversity occurs and why vegetation continually changes. Chapters 2 to 5 deal individually with each ecosystem component as it relates specifically to the vegetation of New Jersey. In this chapter the total impact of all ecosystem components on vegetation is analyzed to identify the particular factors that cause differentiation of plant habitats in New Jersey. This leads finally to a proposed scheme for the classification of major types of natural vegetation in New Jersey. To provide a basis for understanding what follows, the chapter opens with definitions of a few basic ecological terms.

Plant Habitat, Plant Community, and Vegetation Type

The word "habitat" is used in a variety of ways; plant ecologists generally understand it to mean the place in which one plant or a group of plants live. As physical space, then, a plant habitat has meaning in terms of its particular environmental characteristics such as soil type, soil moisture or drainage, temperature, and other conditions that may be critical for plant existence or growth. The degree of variety of plant habitats in a locality or in a whole state such as New Jersey depends therefore upon the amount of diversity in the environmental conditions within the locale or region. An area having uniform relief and climate and similar soil types and water drainage conditions throughout has little opportunity for variation in plant habitat conditions. On the other hand, in some places within a relatively small area there can be wide variations in relief, soil types, and drainage conditions that create different types of plant habitats. This is the case with New Jersey.

Even though as noted in the first chapter every plant species has its own distinctive tolerance range and requirements for light, water, heat, nutrients, and other resources, almost always in nature there is some overlapping of these tolerance ranges. As a result, within any one habitat plants of different species usually grow together. Generally, ecologists refer to the group of plants that live together as a plant community. A plant community and the animals occupying the same habitat compose what is called the biotic community; thus, the biotic community represents the living part of an ecosystem.

Some ecologists have observed that under somewhat the same habitat conditions within a region as large as a state, or even several or many states, the same plant species, more or less, may occur together as a plant community in more than one place. For example, many of the same plant species growing together on sand dunes of the New Jersey coast also grow together on the coast of Long Island. This has led ecologists to classify vegetation into groups called vegetation types. Simply stated, a vegetation type (or plant community type) is a distinctive group of plant species which may be expected to grow naturally together in more or less the same population proportions under particular habitat conditions.

The natural vegetation of New Jersey, like that of other

regions, may be described in terms of its vegetation types. Before describing the types of vegetation, however, it is necessary to delineate the specific types of distinctive plant habitats that are typically found in the area, based on various combinations of environmental factors. A classification scheme for vegetation types in the state can then be proposed.

Moisture Differentiation of Plant Habitats

The factor most important for differentiation among plant habitats within the state of New Jersey appears to be the amount of water in the soil substrate, the medium in which the plant is rooted. As already noted, plants differ in their requirements for water and in their ability to tolerate extremes, whether excessive water or drought. Some plants can live only in standing water; others, such as desert or sand dune species, can exist only under extremely dry conditions, and still others are best adapted to gradations of soil saturation between these two extremes. As examples, in New Jersey the white cedar tree grows mostly in sites that are covered with standing water for at least part of the year while the pitch pine tree is usually found on very dry, sandy soils.

The amount of water in a plant substrate is controlled by the relief of the land surface, the soil type, the height of the water table, and the exposure of the site to wind and to sun. For example, rainwater rapidly runs off steep slopes, and thus little water penetrates below the surface of the soil. On the other hand, standing water may remain in depressions most of the year. Also, soils vary in their capacity to hold the water needed for plant growth. Rainwater drains quickly through very sandy soils, leaving little moisture in the soil for plant growth. In contrast, soils that contain a large amount of clay usually hold a great deal of water. The amount of moisture in the soil also is affected by the level of the water table. A high water table as found frequently in parts of the Coastal Plain may provide standing water at the level of deep-rooted plants. Finally, the amount of sun and wind received on a particular site influences the soil moisture content. A slope that faces south or is exposed to strong winds will be drier than one of similar soil that faces north or has less wind exposure.

The amount of moisture in soil can be considered as a relative gradient ranging from extremely dry to a more or less permanent standing water condition. Five categories in this gradient are suggested for an initial differentiation of plant habitats in New Jersey. These categories and their assigned place on the moisture gradient scale are depicted in Figure 6-1, and each category is assigned a habitat name. Grading from the wettest soil condition (which ecologists call hydric) to the driest (or xeric) condition, the five habitats are as follows:

Marsh habitats are sites on which there is standing water on the soil surface most of the year or which are subject to flooding regularly year round if only for a few days each month or for some part of each day. Included in this category are the wetland sites of tidal areas and the edges of estuaries, rivers, and ponds. In New Jersey marshes occur in inland areas as well as along the coast.

Lowland habitats are sites on which there is standing water on the soil surface for only certain parts of the year, most often in spring and early summer. For example, in the spring, standing water may accumulate in depressions of land or above the soil surface where the water table level is unusually high. By late summer the water on such sites normally will drop to below the soil surface. On lowland adjoining a river, called a floodplain, the land will be covered with water only occasionally, at times of unusually heavy rains when the river overflows its banks.

Well-drained upland habitats represent an idealized midpoint on the moisture gradient, referred to as mesic. Unlike marshes and lowlands, mesic upland habitats usually have no standing water, but in contrast to drier sites, they retain a good supply of rainwater for use by plants. In relief, mesic sites include land of flat or undulating contour including gentle slopes and hilltops which are not excessively drained, and valleys and ravines which are not as wet as swampy lowlands or floodplains.

Ridgetops and slopes of higher elevation or excessively drained flat habitats are drier sites than mesic uplands. Ridge-

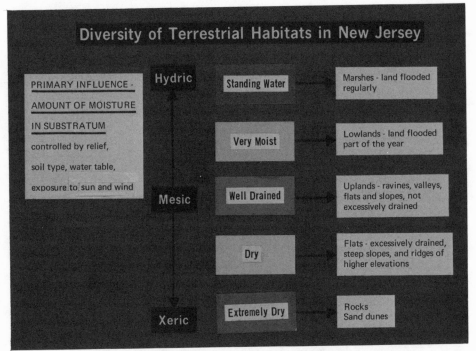

Figure 6-1　Amount of soil moisture as the primary influence causing differentiation of terrestrial plant habitats in New Jersey.

tops and slopes of higher elevations are drier for two reasons: Rainwater is quickly lost because of rapid runoff, and soil water evaporates because of exposure to sun and wind. In New Jersey these conditions are found mostly at higher elevations in the Highlands and the Ridge and Valley sections of New Jersey. The flat sites included in this category are those which are excessively drained primarily because of soil characteristics. For example, rainwater quickly drains through sandy soils which have a limited water-holding capacity, and such soils therefore provide little soil moisture for plant growth. Excessively drained soils occur on the Outer Coastal Plain in the Pine Barrens.

Rock outcroppings and sand dune habitats are sites much drier than those of the previous category and represent the xeric extreme of habitats. In New Jersey the rock outcroppings occur mostly at the higher elevations in the northern part of the state. These are very dry habitats because they lack a soil cover. Just about at the same point on the moisture gradient are the sand dunes of the New Jersey coast on which limited moisture is available for plant growth.

Influence of Climate on Plant Habitats

Variations in several other environmental conditions, while exerting less influence than soil moisture, still have an impact on vegetation in New Jersey. Among these are the variations in temperatures that are described in Chapter 3. The differences in the length of the growing season and the frostfree period particularly appear to be important for plants. The combination of the lower temperatures and shorter growing seasons characteristic of northern New Jersey are more favorable to some plant species than the longer frostfree interval and growing seasons of southern New Jersey, conditions under which other types of plants appear to flourish.

It is not possible to delineate with any degree of scientific exactness a north-south geographic division of New Jersey wherein the temperature conditions on one side favor the group of plants more closely allied with northern regions over those with a more southern alliance. This is because soil and local relief conditions act together with climatic factors to produce conditions more or less favorable to northern or southern plants. However, for the purpose of describing the environmental differences that do occur in the state, a distinction is made between plant habitats of "North" and "South" Jersey based on a dividing line which coincides with the boundary that separates the Piedmont from the Inner Coastal Plain (Figure 6-2). This boundary is just south of the north-south dividing line used for the classification of weather data as described in Chapter 3; the latter division follows county rather than physiographic boundaries.

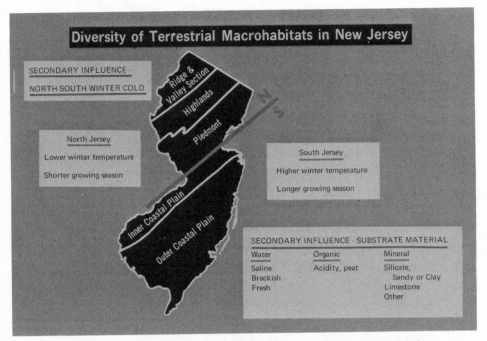

Figure 6-2 Secondary influences causing differentiation of terrestrial plant habitats in New Jersey.

Other Influences on Plant Habitats

In addition to differentiating among plant habitats by the varying amount of soil moisture and by north-south temperature variations, still other habitat characteristics appear to exercise some influence in restraining the successful development of all plants on particular sites. These characteristics relate only to the wetter sites and concern the degree of water salinity and acidity.

Salinity of water. While some plants can successfully grow and reproduce in soil covered or saturated by saltwater, others that flourish under standing freshwater conditions will die if saltwater invades their habitat. Thus, distinctions must be made between saltwater and freshwater habitats. For this purpose, then, in New Jersey the saltwater marshland is placed in a category separate from that of freshwater marshland. The same distinction is not needed for the other lowland habitats — bogs or swamps and floodplains — because these are saturated only with freshwater. Although it may be said that some marshland has brackish water (an arbitrary midpoint between the extremes of freshwater and saline water), for purposes of this book such sites are included in the salt marsh category.

Acidity of water and low substrate fertility. In the above differentiation of habitats on the basis only of soil moisture, one category was designated as lowlands, or sites on which there is standing water over the soil surface for part of the year. This differentiation alone is not sufficient, as it does not define completely enough all the habitat conditions that affect the success of one plant versus another. One such additional condition is acidity.

In some lowlands, because of poor drainage (where there is little or no movement of water) dead plant and other organic remains do not fully decompose to become part of the soil material but rather accumulate to form layers of material called peat. Accumulations of peat are accompanied by very acid conditions and low fertility to which some plants are intolerant. Because of this varying tolerance to the peat condition, further differentiation of wetland habitats is necessary. Lowlands with large accumulations of peat are called bogs and those without large amounts of peat are called swamps or floodplains. As in the case of other classifications, the differentiation between bogs and the other lowlands cannot always be made with scientific exactness, and even ecologists might argue about whether a particular lowland site should be called a bog or a swamp. In later chapters a distinction will be made between the type of plants typically found in a bog as contrasted with those of the fertile and less acid lowlands.

Other environmental differences, while important to some types of plants, appear to be of minor importance in development of a broad classification of plant habitats in New Jersey. For example, it has been reported for other states and appears to be true also of New Jersey that the same type of plant community develops on a variety of soil material. There are some exceptions to plant indifference to the soil types of the state other than the variations in water-holding capabilities. For one, soils derived from limestone material appear to be favorable to more different kinds of plants than any other soil type in the state.

Classification of Terrestrial Plant Habitats in New Jersey

Based on combinations of the environmental factors described above — soil moisture, temperature, water salinity and water acidity — the natural land area of New Jersey can be fitted into a scheme of twelve major types of plant habitats (Figure 6-3):

Type 1 — Saltwater marshes. The salt marshes occur in both North and South Jersey ranging southward from the area of the Passaic and Hackensack river valleys. The marshes are present along the coastal mainland as well as on the bay side of offshore islands and along the southern coast that borders the Delaware Bay and the tidal area of the Delaware River.

Type 2 — Freshwater marshes. While freshwater marshes occur in both North and South Jersey, their areas cannot be clearly delineated on the map because they occur along the edges of lakes, ponds, and rivers and wherever depressions of land are kept flooded on a regular basis by high water tables.

Type 3 — North Jersey bogs. Bogs in North Jersey occur primarily in the area which was once covered by glacial ice. Ice sheets scoured out basins without inlets or outlets which were then filled by water from ice melt. Other bogs were created by glacial deposits which blocked partially or wholly drainage of streams or lakes thus creating poorly drained areas.

Type 4 — South Jersey bogs. Bogs in South Jersey occur especially in stream valleys where groundwater rises to the surface.

Type 5 — North Jersey swamps and floodplains. Swamps occur in glaciated areas of North Jersey particularly in sites of former glacial lakes, for example, the Great Swamp, which is on the site of former Glacial Lake Passaic. Floodplains occur in the broad valleys of the larger rivers such as the Raritan.

Type 6 — South Jersey swamps and floodplains. The swamps

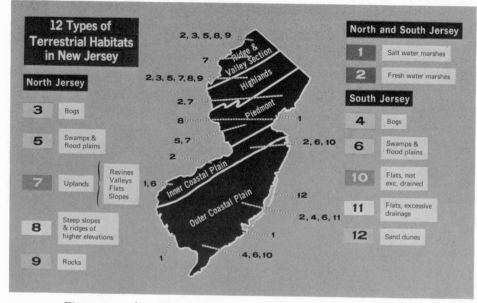

Figure 6-3   Classification of major types of terrestrial plant habitats in New Jersey.

of South Jersey are concentrated along the edges of stream valleys and estuaries. Floodplains also occur in broad valleys of the larger streams such as the upper Millstone.

Type 7—North Jersey uplands. The category of North Jersey uplands includes the slopes, hilltops, valleys, and ravines of the Ridge and Valley and the Highlands sections as well as the flats of the Piedmont.

Type 8—North Jersey ridges and steep slopes of higher elevations. The ridges and slopes of higher elevations occur mainly in the Highlands and the Ridge and Valley sections, though the formations of diabase and basalt on the Piedmont also have examples of the type 8 habitat.

Type 9—North Jersey rock outcroppings. Some parts of the ridgetops in North Jersey are still exposed rock outcroppings, a reminder of the time when the glacial ice stripped off the soil mantle and left the bare rock surfaces exposed.

Type 10—South Jersey uplands. The uplands of South Jersey

consist of the uplands and flats that are not excessively drained. These occur mostly on the Inner Coastal Plain but also include parts of the Outer Coastal Plain that encircle the Pine Barrens (Figure 2-4).

Type 11—South Jersey flats excessively drained. The excessively drained flats of South Jersey comprise most of the region known as the Pine Barrens in the sandy Outer Coastal Plain area. Because of the poor water-holding capacity of the soils in this area, the land is excessively drained and is thus classified as a drier habitat.

Type 12—South Jersey sand dunes. The sand dunes which represent the driest habitats in South Jersey occur along the coast from Sandy Hook to Cape May.

Relationships between Plant Habitats and Vegetation Types

The vegetation of any region becomes the aggregate of plant communities occupying the different types of plant habitats of the area. Thus, in areas undisturbed by man or by natural forces, diversity is dependent solely on the variety of plant habitats in the locale—the more diverse the environmental factors, the more variety in plant life. For this reason if no other, we should expect as many types of vegetation as there are habitats. Thus, the natural vegetation growing on the land of New Jersey can be classified into twelve major types, one vegetation type for each habitat type.

Each of the twelve vegetation types of New Jersey imparts a characteristic appearance because of the growth form of its most abundant (or dominant) species. For example, the characteristic appearance of the North Jersey upland type of vegetation is that of a forest because the most abundant or dominant species are trees. In addition, the larger trees in these forests are mostly tall-growing deciduous broadleaf trees: they lose their leaves in the winter and have a wide leaf compared to the narrow needles found, for example, on a pine tree. The forests of the drier South Jersey Pine Barrens form a sharp contrast in appearance. The forest canopy (the tops of the tallest trees) is not as high and the trees themselves differ in appearance; many are evergreen-needled trees which stay green all winter.

Forests may differ also in the growth that occurs under the tallest trees. They may, or may not, have a lower level of smaller growing (or understory) trees, and in addition, they may, or may not, have a well-developed shrub layer. (By common definition a shrub is a plant with several woody stems, and it usually does not grow as tall as the understory trees.) Finally, plants without woody stems – the herbs – may form an abundant cover on a forest floor or may be conspicuously absent.

Each of the three kinds of wetlands – marshes, bogs, and swamps – has a distinctive appearance also. Since no trees grow in marshes, they have the appearance of grassy fields. Bogs and swamps, on the other hand, do have tree growth but differ in their typical tree species and associated plants. The trees of the New Jersey bogs as classified herein are typically the narrow-needled trees (conifers) as contrasted with the broadleaf trees typical of swamps.

Without the interference of man, the plant communities in natural habitats throughout most of the world would be relatively easy to predict and to describe. But human actions or natural forces as described in Chapter 1 may destroy and modify natural habitats, thereby altering the development of the natural vegetation in the habitat. Therefore, in addition to, or even instead of, having only the "climax" type of vegetation in a particular habitat, various stages of "successional" vegetation may be present. In this sense, then, landscape diversity is dependent on the results of the influence of man and other forces as well as on natural habitat diversity. This is well illustrated in New Jersey, a region long disturbed by man and before that by glacial ice. In each of the twelve basic types of habitats in the state, the vegetation reflects the impact of man's actions as well as of glaciation.

Summary

Despite its small size, New Jersey has diverse vegetation. This is so because of the variety that occurs within the state in landforms, in temperature, in parent rock and soil material, in water drainage patterns, and in water salinity and acidity. Combinations of environmental conditions suggest twelve major types of land habitats and of twelve corresponding vegetation types in New Jersey. Using these types as the framework, the present vegetation of New Jersey is described in the chapters that follow. But since a description of the vegetation of New Jersey based on natural habitat diversity alone would be incomplete, consideration is given also to the results of man's actions on the twelve habitat and vegetation types of New Jersey.

REFERENCES AND SOURCE MATERIAL

Dansereau, Pierre. 1957. Biogeography: An Ecological Perspective. Ronald Press, New York.

McCormick, Jack. 1966. The Life of the Forest. McGraw-Hill, New York.

Odum, Eugene. 1966. Ecology. Holt, Rinehart, and Winston, New York.

Part IV
Vegetation Types of New Jersey

7

Vegetation of the Marshes in New Jersey

Introduction

Marshland includes that part of tidal areas, estuaries, and river and pond edges that is covered with standing water most of the year or that is subject to flooding year-round if only for a few days each month or for some part of each day. Within the category of marshland one more environmental condition controls the type of vegetation that grows in the habitat—the degree of water salinity. Some types of plants, called halophytes by ecologists, are unaffected by salt (sodium chloride) in the water solution absorbed by its roots. Other plants, although able to flourish under standing-water conditions, will die if seawater invades their freshwater habitat. Only a very few plants can bridge the extremes and grow well both in freshwater and in saltwater. For this reason a differentiation is made between two types of marshland habitats—the saltwater marshes and the freshwater marshes.

New Jersey has both saltwater and freshwater marshland habitats. As the starting point of this chapter, the location of the major areas of saltwater and freshwater marshland within the state are identified. Next, the development and change of marshland in time is discussed and, following this, the natural vegeta-

113

tion typical of each type marsh habitat is described. Finally, the impact of man's actions on the marshlands is discussed.

Location of New Jersey Marshlands

The general location of salt marshes is relatively easy to predict, as they must follow the line of coastal tidal waters. However, the type of coastal parent rock formation determines the degree to which coastal marshes develop. For example, there is little tidal marshland along the hard volcanic rock of the Palisades that forms the west bank of the Hudson River. The northern limit of salt marshland in New Jersey starts in the Newark Bay area and extends up the valleys of the Passaic and Hackensack rivers. A huge marshland of about 18,000 acres, known as the Hackensack Meadows, is located in the southern part of the Hackensack Valley. Southward in New Jersey there are salt-water marshes in the Raritan Bay area, particularly where streams empty into the bay. The tidal marshes of the Cheesequake State Park for example are drained and flooded by Cheesequake Creek, which flows into the bay. Small areas of tidal marshland also occur on the inner side of Sandy Hook and along the shores of the Shrewsbury River.

The coastal area from Long Branch to Bay Head has little salt marshland because no offshore island stands in front of the mainland there. From Bay Head southward larger areas of tidal marshes occur. These are present both on the bay sides of the offshore islands and on mainland areas, particularly where streams empty into the bays that separate the mainland from the offshore islands. Around Barnegat and Tuckerton on the mainland there are large stretches of salt marshes and these continue to fringe the mainland coast all the way to Cape May. The Brigantine National Wild Life Refuge, located on an island just north of Atlantic City, has some man-made freshwater marshland in addition to its natural salt marshes.

From Atlantic City southward the area of the tidal marshes grows wider, stretching from the coast of the mainland almost to the bay side of the offshore islands except for the open water areas cleared for the inland waterway.

Tidal marshes also occur along the southern coast of New Jersey in the Delaware Bay area from Cape May up along the Delaware River coast almost to Trenton. Just north of Lower Alloways Creek on the southwestern coast, the marshland is mostly brackish, which is a somewhat arbitrary midpoint between the extremes of fresh and salt water. This is true also of other inland areas of tidal marshes where the seawater is highly diluted by freshwater streams or groundwater. Altogether, New Jersey at present has only about 350 square miles of tidal land; of this, about half, or 177 square miles, edges the Delaware Bay and River; another 140 square miles is along the southern coast of New Jersey in Atlantic, Burlington, and Ocean counties, and the remaining marshland extends from Raritan Bay north to Bergen County.

The areas of natural freshwater marshes cannot be so clearly delineated. This is because this type of marsh occurs wherever depressions of land are kept rather continuously flooded by streams or groundwater. In northern New Jersey the largest areas of freshwater marshes owe their origin to the glaciers. The ice sheets scoured out basins which then were filled by water from glacial melt or by streams whose drainage was altered by glacial deposits. Through the process of succession described in Chapter 1, many of the old glacial lakes have been completely filled by vegetation remains and now are covered with forests so that no one would suspect that a lake once existed on the site. Others remain as lakes, but still others are now marshlands or bogs and swamps; the latter two are discussed in later chapters.

Some part of the area occupied by former Glacial Lake Passaic (Figure 2-3) in Morris County is now the site of the Great Swamp, which contains marsh as well as swamp vegetation. The Hackensack Meadows are on the site of Glacial Lake Hackensack. While the meadows, for the most part, are composed of brackish tidal land, there are also some areas of freshwater wetland. Smaller marshland areas of similar origin occur throughout the glaciated section of northern New Jersey.

In the part of New Jersey not covered by glacial ice, freshwater marshes are found along stream borders and at the mouths of rivers. This is particularly true on the Inner and Outer

Coastal Plains where high water tables keep some of the lower valley land almost continuously flooded. Surprisingly, freshwater marshes, though small in size, also can be found on the driest type of habitat in South Jersey—the sand dunes. Here, freshwater or brackish wetlands sometimes occur in hollows between the dune ridges.

Development and Change of Marshland in Time

In a recent book on the salt marshes of eastern North America, an oceanographer and his wife, John and Mildred Teal, describe the processes by which the salt marshes in stabilized coastline situations may develop and change through time. Shallow parts of a bay may develop into a salt marsh because of the encroachment of marsh vegetation into the bay water by a natural process; stems and roots of the marsh grasses that live at the water edge slow the tidal currents and cause their sediments of sand and organic or inorganic materials to be deposited around the grasses. In addition, as marsh grasses die, their remains add to the deposits. As the fill accumulates, the roots of plants growing at the water's edge can extend outward. Thus, in a continuing process the edge of the marshland creeps farther and farther out. Eventually, sediments and marsh grasses will fill the bays between the mainland and the offshore islands.

As the marshland extends itself out into the bay, the area of land covered by tidal water changes and areas which originally were flooded are left farther and farther from the water's edge. Soon the tides no longer flood these areas; the waters first become brackish as the flooding becomes less frequent. And if the land is never inundated by storm tides, the site may evolve into a freshwater marsh or a swamp. The extension of the marshland into a bay area and the change of saltwater marshes to freshwater marshes and, finally, to a swamp area are additional examples of the process of vegetation succession defined and described in Chapter 1.

All that has just been described does not take place if the coastline is unstable—a condition caused by change in sea level or in the land relative to the sea. It is more common now in our New Jersey salt marshes to find the rising sea level causing an erosion of the outermost margins of the existing marshes rather than to find the marshes encroaching out into the bay. At the time of the maximum glaciation, the coastline of New Jersey probably extended out about 80 miles east of its present line. But as mentioned earlier, there has been a gradual rise in sea level because of the melting of glacial ice. Therefore, where the rise in water level has not been offset by an uplift in coastal land, the sea has encroached landward. In this case, examples of which can be seen in New Jersey, the outer edges of the salt marshlands are being eroded away as the sea encroaches onto the land. This is true of the marshes at Tuckerton. Also areas previously covered by forests may now be salt marshes. Such is the case of large parts of the Hackensack Meadows where large stumps and logs, the remains of a cedar forest, lie buried under the present salt marsh. Ecologists also have found evidence that forest vegetation once existed in the salt marsh area of the Great Bay where the lagoon separates Brigantine Island from the mainland.

A freshwater marshland will develop and change in time much the same way as saltwater bay marsh on a stabilized coastline. In the shallower waters of lakes and ponds, tiny free-floating plants such as the round duckweed may cover large areas of shallow water. Other unrooted plants such as the bladderwort live submerged beneath the surface of the water. Still other aquatic plants live close enough to shore so that their roots penetrate the bottom of the pond or lake. In some of these plants, such as fanwort, the leaves or flowers stay submerged but in others they may emerge above the water's surface as happens with the bulrushes or water lilies whose showy flowers float on the surface (Figure 7-1). Just as the tidal marshlands between the marshland and offshore islands fill in with sediments and the remains of dead plants, the edges of shallow ponds and lakes gradually fill in with debris; the water becomes shallower and a freshwater marsh habitat is created. But, as we will see later, inevitably the marsh habitat in turn gives way in succession to swampland.

Figure 7-1   Floating water lilies on stream in Bass River State Park are an indication of an early stage of the plant succession process in water.

Vegetation of the Salt Marshes (Vegetation Type 1)

The marshes of the New Jersey tidal areas are dominated by two types of grasses, which are the most abundant plants in East Coast salt marshes from New England to the Carolinas. Both grasses have the same Latin genus name, *Spartina,* but their species names differ. One is known as *Spartina alterniflora,* commonly called salt-marsh cordgrass or thatch grass; the other of the two is *Spartina patens,* known as salt-meadow or salt-hay grass. The salt-marsh cordgrass is a coarse, stiff grass as compared with the slender, more delicate salt-meadow grass (Figure 7-2). Early settlers used the salt-meadow grass as feed for their livestock.

Salt-marsh cordgrass, in its tallest form (3 to 10 feet high), normally occupies the marshland zone next to the water, the area that is flooded by high tides twice each day. In contrast, the salt-meadow grass, which is usually no more than 2 feet tall, grows typically in the marshland zone behind the cordgrass farther away from the water's edge, which is flooded only at the higher tides. Both grasses are green in summer and brown in winter and both have special mechanisms for keeping the salts of the tidal water out of the cell sap. The two belts of grasses often form distinct zones on the tidal shores of New Jersey and on other East Coast shores.

A third zone of vegetation occurs on still higher marshland, which is protected from daily tidal inflows and thus may be flooded only monthly at times of very high tides. A grasslike plant called black grass or black-marsh grass which grows only about 1 foot high sometimes occupies large areas of this zone. Its name is somewhat misleading for its color is dark green in summer. Other less abundant plants grow among the black grass on the higher marshland. These include a taller-growing plant called spike grass that looks something like salt-meadow grass but has a spire of tiny flowers. Small shallow depressions called pannes occur commonly on this section of the marshland; they have very saline conditions because they are flooded only infrequently and then remain dry for extended periods during which the water evaporates, leaving the salt behind (Figure 7-2). Particular plants appear to be well adapted for this condition; among these is a plant growing about 1 foot tall called glasswort or samphire, which has swollen green stems that turn bright red in the fall making the marsh almost as colorful as some of the inland forests. Other of the plants commonly found in the depressions include the sea blite, saltwort, orache (or spearscale) and the marsh fleabane.

Above, a small shallow depression (a panne) filled with glasswort contrasts with the surrounding salt-meadow grass in the Tuckerton marsh.

Figure 7-2   Salt marsh vegetation.

The two common species of *Spartina* seen in New Jersey. *Top,* the coarse, stiff, upright salt-marsh cordgrass (*Spartina alterniflora*) on bay side of Sandy Hook. *Bottom,* the salt-meadow grass (*Spartina patens*) has more slender, weaker stems and often cannot hold itself upright as illustrated. Picture also at Sandy Hook.

On higher ground less susceptible to flooding are additional salt-marsh plants, some with more showy flowers. These include a salt-marsh aster, the sea lavender, the marsh (or rose) mallow with hollyhocklike flowers, and the seaside goldenrod and two low growing shrubs, the sea myrtle and marsh elder.

Salt-marsh vegetation with more or less the same zonation and plant composition can be found along most of the eastern and southern coastal marshlands of New Jersey. In the more inland areas where the water tends to be less salty, as in a large part of the Hackensack Meadows, the vegetation is not quite the same. In brackish water some plants more typical of the freshwater marshes join those of the saltwater wetlands. In the mixture either the salt-meadow grass or the freshwater reed grass may be the dominant and most abundant plant growing closest to the water. In the brackish Hackensack Meadows, the reed grass is most successful, but farther south reed grass is not as common in brackish water as salt-meadow or black grass. In both cases growing with the dominant plant are other plants typical of the freshwater and saltwater wetlands (Figure 7-3).

Vegetation of Freshwater Marshes
(Vegetation Type 2)

The particular grasses so abundant in the salt marsh are absent from the freshwater wetlands. Even when former salt marshland changes to freshwater wetlands through the process of succession, the salt grasses originally occupying the site will be replaced by a different group of plants. The same plant species more or less occur together in the freshwater marshlands throughout New Jersey, though their proportions may change particularly with respect to the dominant or most abundant species.

Marshland vegetation as described herein excludes consideration of the aquatic plants that occur in open water areas, some of which are mentioned earlier in this chapter. In the zone of freshwater marshland next to the open water of lakes, ponds, or rivers in New Jersey the plant community typically is domi-

Marsh Habitats in New Jersey

Mostly herbaceous plants grow in both saltwater and freshwater marshes
giving a grassy-meadow appearance to the marshland

Location	Plants of Saltwater Marshes	Plants of Freshwater Marshes
Zone Closest to Water	Salt-marsh cordgrass	Cattail or Reed grass or Wild rice
Inner Zones	Salt-meadow grass Black marsh grass Spike grass Glasswort ⎤ Sea blite ⎟ In saline Marsh fleabane ⎬ depressions Orache ⎟ Saltwort ⎦ Sea lavender ⎤ Salt-marsh aster ⎟ Marsh mallow ⎬ On higher Seaside goldenrod ⎟ land Sea myrtle (shrub) ⎟ Marsh elder (shrub) ⎦	Bulrush Swamp loosestrife Arrowhead Arrow-arum Blue flag Spike rush Bur reed Water dock Sedges Marsh fern ⎤ On higher Swamp milkweed ⎬ land Touch-me-not ⎦

Man's actions that affect marsh vegetation

Fill ⎤
Dredge ⎬ Destroy natural vegetation and reduce plant diversity
Pollute ⎦

Figure 7-3 The Vegetation of Marsh Habitats in New Jersey.

nated by one of three plant species—reed grass, the cattail, or
wild rice (Figure 7-4).

Reed grass, also commonly known by its scientific name—
Phragmites, covers extensive areas in both brackish and freshwater areas of the Hackensack Meadows. This plant grows as
tall as 10 feet, and its showy seed plumes are very conspicuous
in late summer, autumn, and winter; it is the dominant plant
seen when riding from Newark to New York.

A marshland dominated by cattails. Picture taken on Inner Coastal Plain near Cranbury.

Figure 7-4   The three plants that dominate freshwater marshes in New Jersey.

Top, wild rice growing along Rancocas Creek. Picture taken in late September when distinctive seeds were no longer on plant. *Bottom,* reed grass growing in the freshwater (and brackish) marshland of the Hackensack Meadows. Note height of grass relative to person standing in marsh.

In some parts of the Hackensack Meadows and in other freshwater marshes such as the Troy Meadows in Morris County, reed grass is not as abundant as the cattails. Two species of cattails grow in New Jersey: one is narrow-leaved and the other broad-leaved. They occur either separately or together in freshwater marshes and occasionally in brackish wetland. The broad-leaved cattail grows 6 to 8 feet tall and the narrow-leaved is somewhat shorter.

Until recently wild rice was the dominant marsh plant in many freshwater marshlands along the Maurice, Cohansey and smaller rivers flowing into the Delaware Bay and along the Delaware River such as where the Alloway Creek and Salem Rivers flow into the Delaware in the southwestern corner of the state and northward at the mouths of Raccoon Creek at Bridgeport, Mantua Creek upstream from Paulsboro, Woodbury Creek at National Park, Pennsauken Creek at Palmyra, Rancocas Creek between Riverside and Delanco, and in the Trenton marshes. Pollution, drainage and fill-in have destroyed many of these marshlands and stands of the wild rice, an annual plant with stems as tall as 9 feet and distinctive seeds (Figure 7-4), are no longer so abundant in New Jersey. The State Department of Environmental Protection is now in the process of mapping all marshlands and identifying the vegetation of each, but it is known that small patches of wild rice occur as far north as Troy Meadows and good stands of it still can be found growing as far north as the Rancocas Creek as well as in the southern counties of Salem, Cumberland, and Cape May.

Other plants are associated with these three dominant marshland plants but these grow mostly in a zone of the freshwater marshland farther from the water's edge. Most of these are more or less typical of all freshwater marshlands of New Jersey but may vary in abundance depending upon locale and degree of pollution. Reed grass appears to be particularly abundant in polluted marsh areas where little else grows. In a natural undisturbed freshwater marsh there will be low-growing grasslike sedges which form small tussocks above the water level and other plants whose stems may be practically submerged—the bulrush, arrowhead (so named for its leaf shape), swamp loose-strife, arrow-arum, blue flag, spike rush, bur reed, and water dock. On slightly higher land are the marsh fern, the orange-flowered touch-me-not, and the swamp milkweed, which is visited by thousands of butterflies. Beyond this zone of freshwater marshland are the taller woody plants of the swamp or bog habitats. These are the plants that gradually will replace the plants of the freshwater marshland. These processes of vegetation succession or change are described in Chapters 8 and 9.

Man and the Marshland Vegetation

Only recently has there been widespread public interest in preservation of marshlands. This concern is typified by an article on the front page of *The New York Times* on July 13, 1970, entitled "Citizens on U.S. Coasts Rally to Save Tidal Marshes." Yet for centuries, without consideration of the consequences, man has been destroying the natural tidal and freshwater marshlands by drainage, by fill-in, and by pollution. Why the change in attitude?

It is not simply a will to preserve the beauty of marshlands that has stimulated public interest although to many people natural beauty in itself is a value to be cherished. More important is the growing recognition that the wetlands perform functions indispensable to man. The first of these functions, and perhaps the most important, is the unique role that marshlands play in man's food supply.

The U.S. Bureau of Sport Fisheries and Wildlife has estimated that two out of every three species of useful Atlantic fish depend in some way upon the tidal marshlands for their survival. The marshes provide the shelters needed as "nursery areas" for the young of many species. Without the marshes, the fish, clams, shrimp, oysters, and other shellfish on which we depend could not exist. In addition, the marshes furnish homes to a multitude of land-inhabiting animals—raccoons, otters, mink, muskrats, and rabbits among others. Ducks, pheasants, geese, and many other waterfowl use the marshes for nesting or migratory resting places. The fascinating world of marsh life and the food-chain relationships so important to man are described in the book by the Teals that was mentioned earlier and in another book entitled *The Life of the Marsh,* by William Niering.

Other functions performed by marshland are related to flood control and water storage. The coastal and estuary marshlands take the brunt of storm waves, thereby protecting the inner uplands from flooding and erosion. In addition, of the water trapped in freshwater marshes some sinks down to the groundwater table level and becomes part of our drinking water supply. Thus, elimination of marsh areas may exaggerate an already critical problem of short water supply. These important benefits of marshland need not be accompanied by offsetting disadvantages to man; mosquitoes can be controlled by artificial change in water levels during the breeding season or by insecticides not damaging to other life.

The marshland in New Jersey, like that of other states, has suffered severely from man's actions. Of the 12,541 square miles of salt marshes that once existed along the eastern edge of the United States, it is estimated that 50 percent have been destroyed and these just in the last two centuries. One has only to travel through the Newark Bay area to observe the actions that have caused such destruction—drainage, fill-in, and development of the land for industry, housing, or highways, pollution from industries, oil spills, or urban sewage. It is estimated that in recent years the Hackensack Meadows have been filled with waste at the rate of 30,000 tons a week, and plans for the complete development of this area, now valued at more than $100,000 an acre, are under way. As a result of the massive interference with the ecosystem, the variety of plant and animal life in the Hackensack Meadows has been severely depleted. Few other plant species grow with the dominant reed grass and the marsh is lacking in the usual rich assortment of razor clams, mussels, oysters, periwinkles, barnacles, crabs, and snails.

Tidal marshes farther south have been filled for development of seashore houses or for industrial use. Little marshland remains unpolluted in the State and Raritan Bay oysters, once well known in New York restaurants, no longer appear on menus. Most of the oysters have vanished from the bay and those remaining are unsafe to eat.

Freshwater marshes in inland areas have also been claimed for industrial, residential, or agricultural use. Wetlands not so used are being polluted by industrial or sewage water or by excessive use of fertilizers and pesticides. If not completely destroyed by these actions, the vegetation of the marshlands is being so altered that it no longer may provide the habitat conditions needed to support its varied and abundant animal, fish, and waterfowl populations. Fortunately, the New Jersey legislature has recently enacted laws designed to prevent the destruction of the small amount of saltwater marshland that remains.

Summary

The landscape of New Jersey is made varied by coastal tidal marshes, freshwater marshes in the glaciated north, and freshwater and tidal marshes or river edges, estuaries, and lowlands in the south. The vegetation of the tidal marshes differs from that of the freshwater because of varying tolerance of plants to water salinity. Within both types of wetlands, zonations of vegetation occur because different plants are best adapted to particular conditions of flooding and salinity of water.

Natural changes occur in marshland as plants at the water's edge gradually encroach upon the water. The plants of the marshland extend outward and other types of vegetation move in to occupy their original sites which no longer are continually flooded. This is just one example of succession in vegetation.

Marshlands not only serve man as a peaceful refuge from the turmoil of urbanized life but also serve as habitats for a multitude of living organisms. As barriers to storm waves and as water storage areas, the wetlands also serve other functions important to man. For these reasons, it is unfortunate that at an accelerating pace man in New Jersey as in other areas is despoiling the dwindling supply of marshland.

REFERENCES AND SOURCE MATERIAL

Dadario, J. J. 1961. A Lagoon Deposit Profile Near Atlantic City, New Jersey. Bulletin of the New Jersey Academy of Science 6.

Good, R. E. 1965. Salt Marsh Vegetation, Cape May, New Jersey. Bulletin of the New Jersey Academy of Science 10(1): 1–11.

Harshberger, J. W. 1909. Vegetation of the Salt Marshes and of the Salt and Fresh Water Ponds of Northern Coastal New Jersey. Proceedings of the Academy of Philadelphia 61: 373–400.

Heusser, C. J. 1949. History of an Estuarine Bog at Secaucus, New Jersey. Bulletin of the Torrey Club 76: 385–406.

Jervis, R. A. 1963. The Vascular Plants and Plant Communities of Troy Meadows. Bulletin of the New Jersey Academy of Science 8: 1–21.

Niering, William A. 1966. The Life of the Marsh. McGraw-Hill, Inc., New York.

Rosenwinkel, E. R. 1964. Vegetational History of a New Jersey Tidal Marsh, Bog and Vicinity. Bulletin of the New Jersey Academy of Science 9: 1–20.

Teal, John and Mildred. 1969. Life and Death of the Salt Marsh. Little, Brown and Co., Boston.

8

Vegetation of the Bogs
in New Jersey

Introduction

Bogs are a delight to ecology teachers for they not only serve as a home for many unusual plants but also illustrate well the principle and the processes of vegetation succession. New Jersey is fortunate in that natural bogs in varying stages of development still exist in the state. Bogs of different origin occur in North and South Jersey, the geographic division of the state that is delineated in Figure 6-2. Before presenting a description of the vegetation, the formation and stages of bog development in New Jersey are outlined. The chapter concludes with a discussion of the impact of man on the natural vegetation of bogs.

Bog Formation and Development

A bog is a type of lowland (or wetland) site different from marshes, swamps and floodplains. In bog habitats there is not the regularity of flooding that characterizes marshes (Chapter 7), and the substrate conditions in bogs are quite different from those in swamps and floodplains (Chapter 9). The key descriptive terms for the bog habitat are "extreme acidity" and "low fertility." Unlike the lowlands of swamps and floodplains, bogs are wet sites

131

with very poor drainage in which large amounts of organic material such as the remains of dead plants accumulate and do not fully decay. The substrate is one of low fertility combined with extreme water acidity. The latter is important because many microorganisms which normally perform the function of decomposition or breakdown of organic material cannot live under very acid conditions.

Peat is the name given to accumulations of partially decomposed organic material. While better-drained swamps may have some peaty material, bogs are distinguished by very heavy deposits of peat which create the extreme conditions of acidity to which many plants as well as microorganisms are intolerant. Peat is responsible for the dark coffee color of bog water.

The distinction between a bog and a swamp habitat cannot always be clearly delineated, and even ecologists may differ in their interpretation. The authors suggest that a bog as such can be readily recognized by its distinctive vegetation. As described later in this chapter, the physical and chemical environment results in a plant association in which sphagnum moss, heath plants (see Appendix III, Part A) and their close relatives, and certain conifer trees play a predominant role.

Bogs exist in particular areas of northern and southern New Jersey. In the north, the bogs owe their origin to glaciers and thus are found primarily in the parts of the Ridge and Valley, Highlands, and Piedmont sections that were covered by glacial ice (Figure 2-4). There were several ways in which glaciers caused the formation of bogs. The ice sheets scoured out depressions which were then filled by water from ice melt or from streams whose drainage was blocked by glacial deposits. This left deep basins of water often without inlets or outlets. In other cases, huge chunks of ice were buried under sands and gravels. As temperatures gradually rose, the ice melted and left in its place a deep pond (called a kettle hole), again a basin without any drainage channels.

Bogs have formed in southern New Jersey—the Coastal Plains area, beyond the reach of glacial ice—for different reasons. Where the level of the water table is high groundwater may rise to the surface, particularly in stream valleys. The inadequate drainage combined with acid water conditions derived from soils of the Pine Barrens results in heavy accumulations of peat, and bogs are formed. Bogs in southern New Jersey occur mostly as narrow belts along the streams in the Pine Barrens region and on the Cape May peninsula.

No matter what its origin, a bog in the northeastern part of the country will gradually change in time from an open body of water or from a wet lowland into a surface covered with a dense growth of vegetation. So complete is the change that one coming upon the area for the first time could hardly believe that the site once was a wetland.

The gradual change from open water to dense vegetation is just another example of the natural process of succession. The successional process by which marshland plants at the water's edge encroach outward on tidal or freshwater pond areas was described in Chapter 7. The extension of the marshland vegetation in these cases is accomplished primarily by sedimentation, the accumulation of sand and debris that fill shallow water areas so they become exposed land surfaces. In this way, the shore edges gradually converge.

Bogs which initially are bodies of water with relatively steep sides may fill up with vegetation in a manner different from that described for marshland. The process starts at the water's edge where stems of certain land plants extend out over the water and settle on the surface, forming a floating branched network. One such plant that grows in bogs of New Jersey is called swamp loosestrife (or water willow). Several low-growing plants, particularly a type of moss known as sphagnum, quickly fill up the spaces on the floating network of loosestrife branches, and a floating mat of vegetation is developed. Other plants join the initial pioneers and the floating mat gets thicker and thicker and extends farther outward on the water. Even shrubs and small trees start to grow on the thicker portions of the mat. As plant life expands on top of the mat, vegetative parts on the underside of the mat die, drop to the bottom of the basin, and start to fill up the basin along with dead remains of aquatic plants such as water

Figure 8-1   Bog succession in the Pine Barrens at Lakehurst.

lilies. This is why peat deposits accumulate in the bog. Without decomposition action, and particularly in those depressions with poor drainage, the peat layers get thicker and the bog water more acid.

Thus, as the floating mat of vegetation spreads over the water surface, the water basin, which gradually is being filled with organic material, gets smaller and smaller. With the passage of enough time, the floating mat covers the entire water surface, the basin is completely filled with peat deposits, and a forest grows over an area that initially had contained an open basin of water. A few of the bog successional stages that can be seen in New Jersey today are shown in Figures 8-1, 8-3 and 8-4.

Vegetation of the New Jersey Bogs
(Vegetation Types 3 and 4)

Although some of the plants of the bogs in northern New Jersey differ from those in the south, the initial pioneers may be the same, and bogs in both areas show more or less the same characteristic zonation of bog vegetation. In both, an outer zone of sphagnum and shrubs on a floating mat usually, but not always, surrounds the open water area. Farther away from the water's edge is a zone of shrubs and trees. The vegetation of each zone and the differences between the northern and southern bog areas are outlined in Figure 8-2 and described below.

Plants of the Outer Zone Nearest the Water

Most of the so-defined bog sites particularly in northern New Jersey develop with some area of floating mat vegetation, and where this occurs in both northern and southern New Jersey bogs, the outermost portion of the mat is occupied mostly by sphagnum moss and low-growing grasslike plants called sedges. Also at the water's edge may be the purple-flowered swamp loosestrife plant whose stems provide the floating framework to hold the sphagnum and sedges. The leaves of sphagnum moss can hold a great amount of water which imparts a springy cushion-like character to the floating mat. The edge of the mat nearest the open water may not be strong enough to support a man's weight, but it is possible to walk on the mat almost to the edge. Wet feet result, however, for with each step the sphagnum cushion may be depressed below the surface of the water. It is the cushionlike nature of the floating mat that gives rise to the term "quaking bog."

Growing with the sphagnum and sedges in both northern and southern New Jersey bogs is a group of very acid-tolerant shrubs, the most abundant of which include the leatherleaf (named for the texture of its leaves), sheep laurel, the swamp azalea and sweet pepperbush. Wild plants of cranberry, blueberry, and huckleberry also are very common. Less abundant but typical of New Jersey bogs are the shrubs of black alder, stagger-bush, fetterbush, and inkberry. In addition to these, but found only in North Jersey, is a shrub with bluish stems called bog rosemary and another called Labrador tea, both more typical of regions to the north of this state. Large clumps of the rosebay rhodo-

Bog Habitats in New Jersey

Both North Jersey and South Jersey bogs have shrub and forest vegetation
with or without a sedge and sphagnum mat

Com- munity Structure	Plants of North Jersey Bogs		Plants of South Jersey Bogs
Trees	Red maple Black gum Black spruce Larch	Also some Hemlock White pine Yellow birch Southern white cedar	Southern white cedar – the dominant tree and seedlings of Red maple (variety 3-lobed) Black gum Sweet bay
Shrubs	Typically heath shrubs Leatherleaf Laurel Swamp azalea Cranberry Blueberry Huckleberry Bog rosemary Labrador tea Other heaths		Typically heath shrubs Leatherleaf Laurel Swamp azalea Cranberry Blueberry Huckleberry Other heaths Wax myrtles
Herbs	Sphagnum moss Sedges Swamp loosestrife Pitcher plant Sundews Marsh, chain & other ferns Many other herbs		Sphagnum moss Sedges Swamp loosestrife Pitcher plant Sundews Marsh, chain & other ferns Curly grass fern Bog asphodel Many other herbs

Man's actions that affect vegetation

Clear cut ⎫
Fill   ⎬ Destroy vegetation or modify composition
Drain   ⎪
Pollute ⎭

Figure 8-2   The Vegetation of Bog Habitats in New Jersey.

dendron are also found in mature northern bog forests. On the other hand, several species of the wax myrtle shrubs are frequently found in bog areas of the Pine Barrens and Cape May peninsula but not usually in northern bogs.

Plants of the More Inland Shrub-Tree Zone of the Bog

Away from the water's edge more varied plant life is found in both the northern and southern bogs. In this zone the ground is uneven and sphagnum hummocks rise as high as 2 feet above water level. Water still stands in depressions among the hummocks, but in prolonged dry periods the depressions may be dry.

It is on the hummocks for the most part that trees grow in the bog. Also occupying the hummocks at the base of the trees is a strange group of low-growing plants that actually consume small insects and for this reason are called carnivorous plants. Several types of carnivorous plants can be found in almost all the bogs of New Jersey, including the pitcher plant and several species of sundew. Many ferns often grow in the same area, including the marsh fern, cinnamon fern, bracken fern, climbing fern, and chain fern. Also different species of wild orchids frequently can be found on the bog hummocks. One of these, known as the snowy orchid, is said to grow only as far north as the Bennett bog on the Cape May peninsula. Various species of yellow-eyed grass typical of southern areas of the United States occur in New Jersey but only in the southern bogs. Perhaps the most unusual plant of New Jersey bogs is a tiny rare fern known as the curly-grass fern and in this state it is found only in bogs of the Pine Barrens. Other Pine Barren bog plants of interest because of their rarity in this region include the bog asphodel, the slender Carolina club moss, golden-crest, ten-angled pipewort, and species of milkwort, pogonias, and lobelias. In addition to these, the varied species of grasses, rushes, sedges, and many other herbs make the bog a center of botanical interest.

The shrubs of the bog shrub-tree zone are much the same in northern and southern bogs and repeat those found in the outer mat zone. The differences between the northern and southern bogs are most pronounced in tree species.

Tree Species Typical of Bogs in North Jersey

Two trees, the black spruce and the tamarack (or larch), particularly characteristic of wetlands in northern New England and Canada grow on the sphagnum bog mats in northern New Jersey but not on those of the south. Both trees are cone bearing (coniferous) trees with needles as leaves. The tamarack has needles that are light green in summer and tan in autumn and it is one of the few coniferous trees that lose all its needles in winter. Along with these two trees the northern bogs have scattered representatives of hemlock, red maple, yellow birch, white pine, black gum, and alder trees and southern white cedar trees. In appearance, the tree growth in the northern bogs initially is spotty with trees growing just occasionally among the shrubs. Examples of various stages of the North Jersey type of bog growth can be seen at Stokes Forest and at High Point State Park (Figure 8-3).

Tree Species Typical of Bogs in South
Jersey – The Cedar Bog

The early woodland in the bogs of South Jersey presents a sharp contrast to that of bogs in the northern part of the state. The predominant tree is the southern white cedar and it grows in almost pure stands (Figure 8-4). Because of the predominance of the cedar tree certain bogs in South Jersey are called cedar bogs (or cedar swamps) though representatives of the bog shrubs and herbs of the types described earlier also occur on hummocks throughout the bog.

Though narrow in shape, the white cedars grow tall and straight and so close together on the hummocks that they form almost a solid canopy. The dense growth not only makes passage difficult through a cedar bog but also acts to keep the bog cooler in summer and warmer in winter than surrounding open areas.

At left, Cedar bog at High Point State Park. Southern white cedar trees in background; rhododendron grows along trail. *Below,* bog at Lost Lake near Stokes State Forest as seen in the fall of the year. In foreground is floating sphagnum-sedge mat of vegetation. Growing on the other side of the lake are larch trees (light autumn foliage) and black spruce (dark foliage), both northern needled-tree species.

Figure 8-3 Bogs in North Jersey.

Above, aerial photo of cedar bog at MacDonald's Branch showing the typical dense growth of the southern white cedar trees. *Photograph by Jack McCormick. Below,* inside the cedar bog at MacDonald's Branch showing dense growth of cedars that creates a very shaded bog interior. *Opposite page,* typical view of cedar bog from roadside in Pine Barrens. *Photograph from N.J. Department of Environmental Protection.*

Figure 8-4   Cedar bogs of South Jersey.

At the beginning of the nineteenth century Alexander Wilson, an ornithologist, described the cedar wilderness as follows:

These swamps are from half a mile to a mile in breadth, and sometimes five or six in length, and appear as if they occupied the former channel of some choked up river stream, lake, or arm of the sea. The appearance they present to a stranger is singular: a front of tall and perfectly straight trunks, rising to the height of fifty or sixty feet without a limb, and crowded in every direction, their tops so closely woven together as to shut out the day, spreading the gloom of a perpetual twilight below. On a nearer approach, they are found to rise out of the water, which, from the impregnation of the fallen leaves and roots of the cedars, is of the color of brandy. Amidst this bottom of congregated springs, the ruins of the former forest lie piled in every state of confusion. The roots, prostrate logs, and in many places, the water, are covered with green, mantling moss, while an undergrowth of laurel, fifteen or twenty feet high, intersects every opening so completely as to render a passage through laborious and harassing beyond description; at every step you either sink to the knees, clamber over fallen timber, squeeze yourself through between the stubborn laurels, or plunge to the middle in ponds made by the uprooting of large trees, and which the moss concealed from observation. In calm weather the silence of death reigns in these dreary regions; a few interrupted rays of light shoot across the gloom; and unless for the occasional hollow screams of the Herons, and the melancholy chirping of one or two species of small birds, all is silence, solitude, and desolation. When a breeze rises, at first it sighs mournfully through the tops; but as the gale increases, the tall mast-like cedars wave like fishing-poles and, rubbing against each other, produce a variety of singular noises that, with the help of a little imagination, resemble shrieks, groans, or the growling of beasts of prey.

Cedar bogs occur mostly on the Outer Coastal Plain in the Pine Barrens area and on the Cape May peninsula. Examples of the vegetation can be seen in Lebanon State Park. If left undisturbed by man or fire, seedlings of other trees start to grow in the cedar bog and as the white cedar trees mature in age, several types of deciduous trees—mostly red maple, black gum, and sweetbay—encroach upon the bog and crowd out the cedar. In time the vegetation on the site changes from a cedar bog to a southern swamp forest type, as described in Chapter 9.

The Ultimate End of Bog Vegetation
in New Jersey

In both northern and southern New Jersey the final end of bogs is inevitable. As the floating sedge mat extends over the entire water surface or as the basin gradually fills up with plant remains, trees continue to invade the area, becoming so abundant as to form a forest. Thus, any evidence of the water basin from which it all started is obliterated. And as the site becomes drier and drier, the plants that flourish under bog conditions give way to those that develop better on less wet sites.

Man and the Vegetation of the Bogs

Probably since the time of man's initial settlement in New Jersey, the native cranberry and blueberry bushes of both northern and southern bogs have served as food. The berries were important foods in Indian diets and even today large quantities of wild blueberries are harvested for domestic use and for sale.

At about the middle of the eighteenth century man started to use the bog and other lowland areas in New Jersey, particularly those in the Pine Barrens, for commercial growing of cranberries. To use a bog site for this purpose it is necessary to destroy all the natural vegetation. By "turfing," a process wherein all the native plants are excavated from the area, dams are constructed, and ditches made to divert the natural drainage and to control flooding. The area is then ready for planting and cultivation of the cranberry crop.

Additional damage to natural bog vegetation comes from people who claim native plants for their own use or for sale to others. Laurel, swamp azalea, inkberry, and other shrubs are removed from bogs and transplanted elsewhere; sphagnum and other mosses are collected in large quantities and sold to florists as packing material. Branches from white cedar trees as well as those from pine and holly trees end up each year in metropolitan areas for sale as Christmas decorations.

Since the time of the settlement of New Jersey by the Europeans, the white cedar forests of southern bogs have been exploited. It is reported that at the time that the first colonists arrived, some of the cedar trees in the bogs of South Jersey were 6 feet in diameter and more than 1,000 years old. Quickly recognizing the value of white cedar wood for shingles and clapboards as well as for ship building, the early settlers in southern New Jersey began to cut the bog woodlands. Portable sawmills were established near the cedar bogs and then moved as soon as the area was completely cut over. This practice continued into the twentieth century.

During the same time white cedar forests also suffered from the effects of fires that swept uncontrolled through the Pine Barrens area. Though bogs at times can serve as firebreaks, when there are heavy winds the flames may cross a bog through the crowns of white cedar trees. Dr. Silas Little of the Northeastern Forest Service, in a study of the white cedar forests of southern New Jersey, stated that since 1700 "fire and cutting have usually worked together in reducing the proportion of white cedar in favor of associated species." The latter includes the red maple, black gum, sweetbay, and gray birch trees. Altogether, man's actions over the last three centuries have resulted in a drastic reduction in the extensive forests of white cedar that once existed on the lowlands of South Jersey. It is estimated that at present less than 50,000 acres of cedar woods remain in New Jersey.

The New Jersey Water Policy Board, which is concerned with the state water reserves, released in 1970 a report outlining a plan for the development of the water resources in the Pine Barrens region. A group of Princeton University students studied the ecological impact that implementation of the plan would have on the white cedar lowlands. In a report of the study results, edited by Thomas Givnish, it was pointed out that the proposed plan for pumping water out of the Pine Barrens would have a significant impact on the vegetation of the area and indeed would endanger the cedar swamp vegetation. The study identified a number of considerations which should be further explored and evaluated before the pumping plan is authorized.

Summary

The vegetation of the bogs in North and South Jersey reflects two basic ecosystem characteristics described in Chapter 1. Bog vegetation varies according to location in the bog, thereby displaying a zonation of plants with respect to the depth of the standing water. The vegetation of bogs also changes with time and in so doing causes changes in the site conditions. Left undisturbed, a water basin may be transformed into a bog with zonations of vegetation and then into a swampy lowland.

Many plants are common to both northern and southern Jersey bogs, but there are distinct differences particularly in the kind of tree species that first inhabit the bogs. Species typical of regions north of New Jersey extend southward beyond their normal range to the bogs in the northern part of the state. Several species of herbs that are common in the southern United States extend northward beyond their normal range to bogs on the New Jersey Coastal Plains.

Man has destroyed much of the original bog vegetation, particularly the cedar bogs of South Jersey. However, examples of bog vegetation in different stages of development still can be seen in New Jersey state parks located in the Ridge and Valley and the Coastal Plains sections.

REFERENCES AND SOURCE MATERIAL

Givnish, Thomas (editor). 1971. A Study of New Jersey Pine Barrens Cedar Swamps. 1971 Report of the Princeton–N.S.F. Cedar Swamp Study Group, Princeton, New Jersey.

Little, S. 1950. Ecology and Silviculture of White Cedar and Associated Hardwoods in Southern New Jersey. Yale University School Forestry Bulletin 56: 1–103.

McCormick, Jack. 1970. The Pine Barrens: A Preliminary Ecological Inventory. New Jersey State Museum Report No. 2. Trenton, N.J.

Montgomery, J. D. 1963. Flora of the Bennett Bog Wildlife Sanctuary, New Jersey Nature News, Audubon Society 18: 111–124.

Montgomery, J. D. and D. E. Fairbrothers. 1963. A Floristic Comparison of the Vascular Plants of Two Sphagnous Wetlands in New Jersey. Bulletin of the Torrey Club 90: 87–99.

Niering, W. A. 1953. The Past and Present Vegetation of High Point State Park, New Jersey. Ecological Monographs 23: 127–148.

Rosenwinkel, E. R. 1964. Vegetational History of a New Jersey Tidal Marsh, Bog and Vicinity. Bulletin of the New Jersey Academy of Science 9: 1–20.

Wilson, Alexander. 1824. American Ornithology. Vol. VIII.

9

Vegetation of Other Lowlands: Swamps and Floodplains in North and South Jersey

Introduction

Swamps and floodplains, like bogs, are lowlands on which standing water is present for only part of the year, most often in spring and late fall. However, because of better drainage conditions, the substrate in swamps and floodplains is not as acid as that of bogs. As a result, neither swamps nor floodplains have the large accumulations of peat that characterize the bog habitat.

Though swamps and floodplains differ in origin, their moisture conditions tend to be similar enough in New Jersey so that the same group of plants grow in both places. But some plant species that grow well in swamps and floodplains in North Jersey are not found in the same type of habitat in South Jersey, and the reverse situation also exists. As described in Chapter 6, differences in the intensity and duration of cold weather combined with soil and drainage variations appear to account for the difference in vegetation between lowlands of North and South Jersey. However, as mentioned in Chapter 6, one must remember that the north-south division of New Jersey used in this book is not exact in the real world. Though usually in sparse numbers, representatives of plants identified as having a more southern

147

affinity may, in fact, be found north of the Coastal Plains, particularly on the Piedmont, and northern species may occasionally extend over the boundary, especially to the Inner Coastal Plain.

Origin and Formation of Swamps and Floodplains

Many of the swamps of North Jersey owe their origin to the glaciers and for this reason occur primarily in the northern part of the Ridge and Valley, Highlands, and Piedmont sections, the areas that were glaciated (Figure 2-3). Since the disappearance of the ice sheets some 12,000 to 15,000 years ago, former glacial lakes or ponds have been edged by marsh vegetation (Chapter 7) or by bog vegetation (Chapter 8).

As marsh plants gradually extend farther and farther into the waters of the ponds or lakes, sediments accumulate in the original marsh-plant sites, which then become drier. The drier conditions favor swamp vegetation. This sequence of events has taken place in areas on the site of the huge Glacial Lake Passaic (Figure 2-3). After glaciation, the lake drained, but there were places of shallow undrained depressions on the lake bed; here the successional process took place. Such is the background of the Great Swamp near Morristown which now is occupied by marsh, swamp vegetation, and drier woodlands. Bogs also may be transformed into swampland as swamp-type plants inevitably replace the bog vegetation. In these cases the swamp habitat results from the successional process of vegetation which is accompanied by the change of a very wet site to one of less wet habitat conditions (Figure 9-1).

The coastal plains of South Jersey were beyond the reach of the glaciers and yet they too have swamps. The topography and drainage conditions in this area account for the lowland conditions. The New Jersey coastal plains for the most part have flat or only gently rolling topography; stream flow is sluggish and no well-cut river valleys exist. In addition, in many areas there are shallow depressions underlaid by clay which act to contain water. These conditions combined with the low elevation and high water tables create flooding above the land surface for at least

Figure 9-1 Marsh and swamp successional process as seen in a lake. Marsh plants have gradually encroached into the middle of Lake Carasaljo in Lakehurst; and at the same time, as sediments fill in the water around the lake perimeter, swamp type vegetation replaces the marsh plants. Someday, unless man interferes, no open water will remain as a lake.

part of the year. Swamp conditions occur throughout the Inner and Outer Coastal Plains but are concentrated particularly along the edges of stream valleys and estuaries. As mentioned in Chapter 8, whether the drainage of a particular site is poor enough to classify the habitat as a "bog" rather than as a "swamp" causes confusion even among ecologists, who sometimes use the words interchangeably. Herein, however, the distinction is made between land where very acid and low fertility conditions favor the development of the characteristic bog vegetation (Chapter 9). Swamps are less acid and more fertile sites which are inhabited by a different type of vegetation.

Floodplains are well-defined, broad, flat, valley surfaces that are covered with water when a stream overflows its banks. The Raritan River from New Brunswick to Bound Brook has a well-formed floodplain as do its northern tributaries such as the Lamington River. Also, the Millstone River, which empties into the Raritan, has a wide floodplain area from Princeton north to its outlet. Each of these floodplains illustrates the classic structure of this type of formation. The land closest to the water (the outer part of the floodplain) is slightly higher in elevation and better drained than the inner or back side of the floodplain. The inner part is not usually as well drained because it is lower and receives seepage from higher adjacent terrace land and because its soil is less sandy texture than that of the outer floodplain. Sometimes subsidiary streams called yazoos may develop on an inner floodplain.

Though the frequency, duration, and depth of flooding of the floodplains may vary, the resulting condition of poorly drained land creates just about the same type of plant habitat as that of swampland of different origin. Thus, one may talk of the "floodplain swamps" particularly when referring to the more poorly drained inner section of the floodplain.

Swampy areas have been both created and later destroyed in New Jersey as in other areas by beavers. Beavers winter in dome-shaped lodges which they build of plant stems, leaves, and mud, and construct dams to make water of slow-flowing streams deep enough so that the entrance to the beaver lodge is beneath the water surface. The impounding of water by beaver dams changes the surrounding land. Existing trees may die because of intolerance to the overflooding conditions. When the beaver family moves on to new locations, its dams gradually erode away and the pond area reverts to a swamp area.

Beavers are not nearly so plentiful now in New Jersey as they once were, but the results of present dam-building activities still can be seen; Flatbrook River in Stokes Forest shows evidence of beaver activities (see Figure 5-4).

Vegetation of Swamps and Floodplains of North Jersey (Vegetation Type 5)

Even within the area designated as North Jersey (the Ridge and Valley, Highlands, and Piedmont regions), slight differences in plant species occur between the more northern lowland habitats and those in the Piedmont section.

At High Point State Forest in the northwest corner of the state two distinct zones of vegetation often can be observed in the swamps. The wettest sites are densely covered by thickets of various types of shrubs. The most abundant of these—the buttonbush, alder, and willow—may form pure stands or may grow together in mixtures. Growing with these species but less abundantly are other shrubs including the swamp azalea, winterberry, several species of viburnum, the highbush blueberry, spicebush, and witch hazel. The less wet swamp sites in the north are covered with forests. Typically, red maple and to a lesser extent yellow birch are the characteristic trees in these swamp forests and in some cases they account for as much as 70 percent of the total tree cover. Other less abundant trees include the black ash, white ash, basswood, tulip tree, black gum (also called sour gum), and the lower-growing hornbeam trees. A variety of shrubs grow underneath the tree canopy and include mostly the species that form the shrub thickets already described. On the floor of the swamp forest grow the herbs—most typically skunk cabbage, cinnamon fern, and a variety of sedges and mosses.

Going southward to the swamps of the Piedmont in central New Jersey a slight change in the swamp forest can be seen. Red maples still are very abundant in central New Jersey, but fewer yellow birch trees occur. Instead, representatives of the elm, pin oak, swamp white oak, and silver maple trees are more abundant, and these join the black gum and ash trees as the most common trees in the swamps of the Piedmont. A mixture of these species can be seen in the swamps occupying the site of former Glacial Lake Passaic, as in Troy Meadows and in the Great Swamp near Morristown. The shrubs typical of swamps in northern New Jersey also occur farther south although they are joined by others; for example, in the Great Swamp the mountain laurel and shadbush shrubs are abundant. Spicebush also is very abundant on moist land of the Piedmont. The herbs of the Piedmont swamp forests are much like those in the Ridge and Valley section and include skunk cabbage, ferns, mosses, and a

Above, swamp forest in springtime on the Raritan River floodplain with a carpet of the early flowering herb—bluebells. *Opposite page,* swamp forest of the Delaware floodplain on Bull's Island that has growth of huge lowland trees. Photograph shows a sycamore tree typically distinguished by its large patches of mottled bark resulting when the outer darker bark peels off exposing the creamy-white inner layer. The open spreading crown also is characteristic of the sycamore.

Figure 9-2   Swamp vegetation of North Jersey.

number of plants that bloom in early spring, such as the marsh marigold, trillium, and several species of orchids.

The greatest variety in the composition of lowland forests occurs on the floodplains of New Jersey. On the better-drained outer sections of the floodplains of the Raritan and Millstone rivers, the characteristic trees include the willow, river birch, sycamore, and box elder (Figure 9-2). On the more poorly drained inner floodplains of these rivers, those tree species typical of the central New Jersey lowlands grow abundantly; these include the ash, pin oak, silver maple, swamp white oak, red maple, elm, and black gum with occasional representatives of the more upland species. The fragrant smelling spicebush is the most common shrub. On the floor of the forest among the remains of fallen trees grow many herbs such as the sensitive fern, touch-me-not, may apple, jack-in-the-pulpit, spring beauties, trout lilies, and the cardinal flower. In areas where forests have been disturbed by man or where trees have been downed by strong winds, vines are common and grow almost to treetops. These include poison ivy, Virginia creeper, Japanese honeysuckle, bittersweet, and wild grape.

On the better-drained portions of the Raritan floodplain west of New Brunswick, trees typical both of swamps and of the upland intermingle, and no one species is so abundant that it can be called dominant. The result is a forest of unusual diversity, with representatives of all the trees mentioned in the preceding paragraph and a few additional tree species such as the Norway maple, honey locust, sassafras, black cherry, sumac, ailanthus, mulberry, and hawthorn. Shrubs include the spicebush, elderberry, several species of viburnum, Japanese barberry, and the silky dogwood. In many places the herb bluebells forms a spring carpet (Figure 9-2).

Vegetation of Swamps and Floodplains of South Jersey (Vegetation Type 6)

The habitat conditions in the lowland swamp and floodplain areas of South Jersey appear to favor the growth of certain tree species that are not commonly found in states north of New Jersey. In this state the species grow mostly on the Coastal Plains and occur only locally on the Piedmont. As a result, the forest composition on the lowlands of South Jersey is quite different from that described for North Jersey (compare Figures 9-2, 9-3, and 9-4). In addition, within South Jersey, differences in soil conditions between the Inner and Outer Coastal Plains and in temperatures between the Cape May peninsula and the more northern Coastal Plains contribute to other distinctions in lowland forest composition.

On the lowlands of the Inner Coastal Plain a unique type of swamp forest develops on wet sites which previously were clearcut or used for agriculture. Pure or nearly pure stands of even-aged sweet gum trees develop on such sites. In New Jersey this tree is mostly confined to the Coastal Plain though scattered representatives also grow in swamp or floodplain areas in the southern part of the Piedmont. As the sweet gum woodland matures, other trees become more abundant. These include two trees of a southern affinity—the willow oak and the Spanish oak—as well as the red maple, pin oak, tulip tree, beech, swamp white oak, ash, elm, and sassafras. An occasional holly or persimmon tree, the latter also typical of forests south of New Jersey, also can be found in the woodlands.

The shrubs growing in the more mature swamp forests of the Inner Coastal Plain include the arrowwood, spicebush, highbush blueberry, sweet pepperbush, and swamp azalea. Poison ivy and honeysuckle are abundant in certain areas, frequently making dense thickets of the lowlands. Wild garlic is a common herb.

The Outer Coastal Plain accounts for more than two-thirds of the land of South Jersey (Figure 2-4). Its swamp and floodplain habitats may be covered with pure growths of the chain fern, shrub thickets, or forests. The natural vegetation succession is from the fern community to a shrub thicket to a white cedar bog forest, and finally to a forest type identified as the southern swamp hardwoods.

Widespread groups of chain fern may occur along the edges of streams or ponds in the Pine Barrens. The shrub thickets

Swamp and Flood Plain

	Both North Jersey and South Jersey Lowlands	
	Plants of North Jersey Lowlands	
Community Structure	*Central N.J.-Piedmont*	*More Northern N.J.*
Typical Trees	Pin oak	Yellow birch
	Red maple	Red maple
	Ash	Ash
	Elm	Basswood
	Swamp white oak	Tulip tree
	Black gum	Black gum
	Silver maple	
	Also on Floodplains	
	Willow	Box elder
	Sycamore	River birch
Typical Shrubs	Spicebush	Alder
	Witch hazel	Willow
	Arrowwood	Buttonbush
	Viburnums	Spicebush
	Others	Witch hazel
		Others
Typical Herbs	Skunk cabbage	Skunk cabbage
	Spring herbs	Spring herbs
	Sedges & mosses	Sedges & mosses

Man's actions that affect vegetation

Figure 9-3   The Vegetation of Swamps

grow in the same type of locale and for the most part these consist or pure stands of the leatherleaf or the highbush blueberry shrub or mixtures of the two with representatives of sweet pepperbush, huckleberry, fetterbush, swamp azalea, and sheep laurel.

Because white cedar trees are not shade-tolerant — that is, their seedlings do not develop well under shady conditions — the white cedar bog forest naturally and inevitably gives way to a

Habitats in New Jersey

are covered with forests or shrub thickets.

Plants of South Jersey Lowlands

Inner Coastal Plain	Outer Coastal Plain
Sweet gum	Swamp hardwoods
Red maple (3-lobed)	Red maple (3-lobed)
Beech	Sweetbay
Willow oak	Black gum
Spanish oak	Some Southern white cedar
Swamp white oak	
Tulip tree	

Also on Cape May

Water oak	Pond Pine
Loblolly pine	Basket oak
Spicebush	Sweet pepperbush
Arrowwood	Blueberry
Blueberry	Swamp azalea
Sweet pepperbush	Leatherleaf
Swamp azalea	Other heaths
Others	
Vines (woody)	Chain fern
Wild garlic	Mosses
	Sedges

Clear cut
Fill } Destroy vegetation or modify composition
Drain

and Floodplains in New Jersey.

swamp forest of tree species that are shade-tolerant. On the Outer
Coastal Plain in South Jersey, this usually is the southern hardwoods swamp forest which consists typically of three principal
trees — the sweetbay (or southern magnolia), a tree not found in
North Jersey; the black gum tree which does occur in northern
lowlands; and the red maple in a variety identified as three-
lobed. (The red maple tree grows in both dry and moist sites of
North Jersey but not as the three-lobed variety.) These three

trees occurring in varying mixtures are the dominant trees of the southern swamp hardwood forests; of the three, red maple is most common and sometimes accounts for as much as 80 percent of the trees making up the forest canopy (Figure 9-4). Though not as abundant, other species of trees also are found in the swamp hardwood forests, including the gray birch, southern white cedar, sweetgum, tulip tree, ash, sassafras, and an occasional American holly and pitch pine, the last two being more characteristic of drier sites. In the lowlands of Cape May, several tree species characteristic of swamp forests in the southeastern United States have been found. These include the loblolly pine, the pond pine, and two species of oak; one is commonly known as the basket (or swamp chestnut) oak tree, and the second is the water oak, which reaches its northern limit in Cape May and southeastern Cumberland County (Figure 9-5). Although Dr. Witmer Stone reported in 1910 that the southern bald cypress tree also was a native to Cape May County, Louis Hand and Robert Alexander in their 1965 survey of Cape May trees state that the reported native specimen of the tree in this county probably was brought to New Jersey from South Carolina.

Growing most abundantly under the trees in the southern hardwood swamps are shrubs of sweet pepperbush, highbush blueberry, and swamp azalea. Also common but not as abundant are other species of blueberry, huckleberry, fetterbush, sheep laurel, and leatherleaf. On the forest floor grow mostly the cinnamon and chain ferns, rushes, sedges, sphagnum, and other mosses.

In some parts of the Pine Barrens, a different type of forest has developed in lowland depressions. It is generally identified as a pine lowland forest type and is composed mostly of pitch pines, the tree more typical of sandy drier soils. The height of the forest depends upon the frequency of fire. Growing with the pine may be scattered individuals of red maple, black gum, and gray birch. The leatherleaf usually forms most of the shrub growth although the shrubs typical of the swamp hardwoods forest also are found in this type of lowland vegetation. This includes sheep laurel, black huckleberry, and dangleberry. Also there is a well-de-

Left, an even-aged lowland stand of sweetgum trees growing on the Coastal Plain near New Lisbon. Star-shaped leaf of this tree makes it easy to distinguish. *Right,* a swamp hardwoods forest typical of South Jersey. Leaves of a sweetbay tree appear in the lower right corner and a sweetgum tree at extreme right; a red maple is the large tree in center, and the large tree at extreme left is a willow oak. Forest also has numerous black gum trees and a few American holly; and blueberry is a common shrub. Photograph taken in Belle Plain State Forest.

Figure 9-4   Swamp vegetation of South Jersey.

veloped lower herbaceous layer in the forest including bracken fern, wintergreen, and the turkeybeard.

Many interesting low growing herbs can be found in the lowlands of the Pine Barrens. Among these are several species of orchids, milkworts, asters, pipeworts, rushes and sedges as well as the golden club, swamp-pink, grass-pink, tawny cotton grass, turkeybeard, yellow-eyed grass, starflower, rose pogonia, golden-crest, arethusa, and bog asphodel.

Man and the Lowland Vegetation

Swamps like marshes can be completely obliterated by draining or by filling the lowlands. By doing both, man has caused the disappearance of many acres of swamp vegetation in New Jersey. It was only because of the persevering efforts of many dedicated people that the Great Swamp near Morristown was saved from conversion into a jetport. Although housing developments and garbage dumps have encroached on much of the original lowland that represents the remains of Glacial Lake Passaic, more than 4,000 acres of swamp have been preserved as a natural wildlife refuge. In addition, swampland is preserved in state parks and forests such as High Point in the Ridge and Valley section and Lebanon in the Outer Coastal Plain.

Against the advice of scientists, the land of the floodplains has been developed for residential or industrial buildings. As predicted, the result has been flooding of this or compensatory land at times when streams are unable to contain heavy downpours of rain or accumulations of snow melt.

On the swamps and floodplains in New Jersey that have not been completely destroyed, there is little virgin vegetation. As early as 1854 it was reported that not a single acre of original growth remained in the swamps of New Jersey and by that time some areas had been cut over a second or third time. Lumbering practices can change the composition of the swamp forests. For example, on lowland sites of the Outer Coastal Plain complete clear cutting tends to favor the development of a white cedar forest stand, while partial cutting of timber on the same site tends to

Figure 9-5 Lowland forest in Cape May. Large tree in center is a pond pine, a typically southern tree which reaches its northern limit in Cape May and Cumberland counties. Picture was taken near South Dennis where this same forest also contains representatives of loblolly pine, another southern tree.

favor development of the swamp hardwood type of lowland forest. This is because the shade-tolerant hardwoods can resprout from stumps, whereas the reestablishment of white cedar depends upon natural reseeding and development of these seedlings is most successful under conditions of full sun. Fire tends to favor the development of the swamp hardwoods at the expense of white cedar again because of the ability of the hardwoods to resprout.

The effects of stream pollution on swamp vegetation are not fully known but there is reason to believe it is destructive. One ecologist, for example, has observed that the number of plant species growing along the banks of the highly polluted Whippany

River is considerably less than that along the edge of Troy Brook which is not inundated with water containing industrial wastes.

As mentioned in Chapter 4, the counties of and around the Pine Barrens which until now have been relatively sparsely populated are increasing their numbers at the fastest rate in the state. In many of the new residential developments, household waste disposal is by individual septic tanks rather than by municipal sewer and waste treatment systems. Because of the sandy nature of the soils in the area and the high groundwater tables, there is increasing danger that the sewage will cause pollution of the waters of the Pine Barrens. This would be extremely serious for two reasons. First, the present unique natural vegetation may be harmed and even destroyed by the pollutants. Second, perhaps of greater concern to all residents in New Jersey, is that pollution of the groundwater in the Pine Barrens area would endanger the present and future water supplies of the state.

The groundwaters of the Barrens have long been considered one of New Jersey's most important resources. This was the supporting reason for the acquisition of the land making up the Wharton State Forest, a tract of almost 100,000 acres. The land was purchased from a Pennsylvanian (Joseph Wharton) who had hoped to tap the groundwater and export it out of the state to supply the needs of the Philadelphia area. The purchase of the land by the New Jersey legislature was an important step; however, with the passage of each year the need to develop a total plan for the utilization of the Pine Barrens land and other resources becomes more pressing. More will be said about this in the last chapter.

Summary

Variations exist in the natural vegetation of the swamp and floodplains of New Jersey. Between the lowlands in North Jersey and those in South Jersey there is considerable difference, particularly in the types of tree species that grow most abundantly in the forests. The boundary between the Piedmont and the Inner Coastal Plain appears to delineate more or less a dividing line between plants of a more northern or more southern affinity. Within the areas designated as North and South Jersey finer differences in habitat conditions cause additional diversity in species composition. Lowland vegetation naturally progresses through successional stages—a shrub thicket develops into a forest and even the forest type may change in time as is the case of the swamp vegetation in South Jersey.

By his actions man has destroyed much of the original natural vegetation of the swamps and floodplains in New Jersey and has modified the composition of the remaining vegetation.

References and Source Material

Bernard, J. M. 1963. Lowland Forest of the Cape May Formation in Southern New Jersey. Bulletin of the New Jersey Academy of Science 8: 1–12.

Buell, M. F. and W. A. Wistendahl. 1955. Flood Plain Forests of the Raritan River. Bulletin of the Torrey Club 82: 463–472.

Hand, Louis and Robert Alexander. 1965. Native Trees and Shrubs of Cape May County, New Jersey. Cape May Geographic Society, Cape May, New Jersey.

Jervis, R. A. 1963. The Vascular Plants and Plant Communities of Troy Meadows. Bulletin of the New Jersey Academy of Science 8: 1–21.

Little, Silas. 1951. Observations on the Minor Vegetation of the Pine Barren Swamps in Southern New Jersey. Bulletin of the Torrey Club 78: 153–160.

McCormick, Jack. 1970. The Pine Barrens: A Preliminary Ecological Inventory. New Jersey State Museum Report No. 2. Trenton, N.J.

Niering, W. A. 1953. The Past and Present Vegetation of High Point State Park, New Jersey. Ecological Monographs 23: 127–148.

VanVechten, G. W., III and M. F. Buell. 1959. The Flood Plain Vegetation of the Millstone River, New Jersey, Bulletin of the Torrey Club 86: 219–227.

Wistendahl, W. A. 1958. The Flood Plain of the Raritan River, New Jersey. Ecological Monographs 82: 129–153.

Zuck, R. K. 1963. The Great Swamp of New Jersey. Garden Journal of New York Botanical Gardens 13: 169, 170, 173.

10

Vegetation of the Mesic Uplands in North Jersey

Introduction

The area designated as North Jersey includes a broad array of site reliefs ranging from the relatively flat and gently rolling lands of the Piedmont to the more rugged Highlands and the Ridge and Valley regions where contrasting high ridges, steep slopes, narrow ravines, and broad valleys make for a diverse landscape (Figure 2-5). Underlying this topography is a variety of parent rock materials—conglomerates, sandstones, shales, limestones, gneiss, and schist as well as volcanic basalt and diabase. In addition, depending upon location, the parent rock formation may, or may not, be covered by glacial deposits of varying age. The result is natural land with soils of varying types, textures, depths, and degree of stoniness.

It is the variations in land relief and in soil conditions with the resulting differences in environmental conditions that create the distinctive plant habitats in North Jersey. Previous chapters deal with the wetter habitats of North Jersey—the marshes, bogs, swamps, and floodplains. In this chapter the natural vegetation of the mesic upland plant habitat of North Jersey is described (Vegetation Type 7); the following chapter is devoted to the vegetation of drier habitats—that of the ridgetops and steeper slopes

164

of high elevations (Vegetation Type 8) and of rock outcroppings, a habitat of extremely variable environmental conditions (Vegetation Type 9).

Habitats classified as mesic uplands represent an idealized midpoint on the soil moisture gradient. Unlike the wetter swamps and other lowlands, there is no standing water on the uplands. But unlike drier sites—for example, those of ridgetops—upland sites retain a good supply of moisture in their soil. Recognize, however, that the category of well-drained uplands in North Jersey includes slopes, hilltops, valleys, and ravines in addition to land of flat topography. Surprisingly, the natural vegetation growing on these diverse landforms and their varied soils might not show much variation today but for the impact of man. This is so because in North Jersey as in many other areas the same group of plants more or less appear to be able to grow well on a wide variety of relief and soil types provided the climatic and soil moisture conditions are about the same.

Because past actions of man have had such a marked influence on the present vegetation of the North Jersey uplands, the chapter starts with a brief review of this impact. This is followed by a delineation and description of the types of plant communities found in the upland habitat. The chapter closes with speculations on what the vegetation of the North Jersey uplands might be but for the presence of man.

Influences on the Upland Vegetation in North Jersey

The influence of man on natural vegetation throughout New Jersey was reviewed in Chapter 4; fire, timbering practices, and the selective use of land for settlement and agriculture have left a major imprint on the present upland vegetation of North Jersey.

Although excessive cutting of the North Jersey woodlands and devastating forest fires have been minimal since the start of the twentieth century, the imprint of past results of the combined factors remains. The practice of cutting woodlands as soon as trees are large enough for cordwood and recurring fire damage both favor certain tree species over others that grow naturally in the upland habitat of North Jersey. This is so because some trees have more ability to resprout either after being damaged by fire or when cut. Pitch pine and oak trees of sapling size (having a diameter of 1 inch at 4 feet above ground) or even smaller usually have the capacity to put forth new trunks and vegetative growth if the existing growth is killed (Figure 1-1). In contrast, other upland tree species such as the sugar maple and hemlock have little capability for resprouting after their trunks are decimated and can perpetuate themselves only through natural seed reproduction. Also, some tree seedlings (trees less than 1 inch in trunk diameter) can survive the heat of fire better than others less tolerant to heat.

In still another way both fire and timbering practices have influenced forest composition in northern New Jersey. The cutting of forests or the regular recurrence of fire can create a more open woodland condition, one in which sunlight penetrates to the floor of the forest. In contrast, in the absence of cutting or of fire, upland trees grow taller and their tops form a closed canopy, shutting out the rays of the sun. Additional shading of the forest floor may come from a dense growth of shrubs which normally would be eliminated by fire. Whether or not sun penetrates to the floor of a forest is important because the conditions of shade are favorable to particular tree species, the shade-tolerant species, but not to many others which propagate or develop well only in sunlight. In North Jersey only the young trees of sugar maple, hemlock, yellow birch, sweet birch, beech, hop hornbeam, and dogwood appear to develop well in full shade; in contrast, seedlings of the oaks, hickories, and the tulip tree grow rather poorly in full shade. In several ways, then, past fires and timbering practices have provided on the North Jersey uplands the conditions favorable for development of certain species, particularly the oaks. In contrast the development of other species such as the sugar maple, hemlock, and birches has been hindered by man's actions. The result is seen in the composition of the present vegetation as discussed later.

Also of importance has been man's selective use of the uplands of North Jersey. The relationship between population and forest land in the whole state of New Jersey was reviewed in Chapter 4 and the data indicate that the counties in the northwestern part of the state have relatively fewer people and more forests per square mile (Figure 4-3) than other parts of New Jersey. But within the area both the population and the woodlands are unevenly distributed. For example, in 1950 Kittatinny Mountain, which accounts for 1.8 percent of the land of New Jersey, was 64 percent wooded and occupied by only 1,700 people. More people, however, live in the adjoining valley where most of the land has been cleared and is used for farming. Except for small woodlots maintained primarily to supply each farm with its fuel wood, fencing material, and timber for construction, little natural vegetation remains on the fertile valley land in North Jersey. The forests that do occur here are found on the less fertile slopes or hilltops or in deep ravines, and even these no longer represent virgin growth. In recent years some of the less fertile land that once was farmed has been abandoned. On such land various stages of successional vegetation can be seen.

Another major influence on the upland vegetation of North Jersey has been that of the chestnut blight discussed in Chapter 5. The state geologist, in his 1899 report on forest conditions, stated that the American chestnut tree was one of the most widely distributed large trees in North Jersey and he noted that on good soil trees ranging from 42 to 60 inches in diameter and from 40 to 60 feet in height were not uncommon. Since then, the top growth of all mature chestnut trees has been killed by disease; while the stump and roots of many chestnuts remain alive and continue to send out sprouts, the sprouts die back after reaching a growth of about 15 feet.

Types of Mesic Upland Plant Communities in North Jersey (Vegetation Type 7)

By some of the actions described above, man has totally destroyed much of the natural vegetation that once existed in northern Jersey; but surprisingly, some of his actions have increased the variety of the vegetation that does remain. For example, if they had been left completely undisturbed, the uplands of North Jersey today would be blanketed only by forests. At present, in addition to forest vegetation, so-called successional (or nonclimax forest) types of plant communities can be found on land that has once been disturbed by man but then left untouched. Furthermore, the upland forests that do exist might be less diverse if man had never settled here. Today three rather different types of forest communities can be found on the mesic uplands of North Jersey, and many ecologists believe that without man's influence only two of these would prevail over most of the uplands. But before discussing this possibility further, the basic types of plant communities presently found on the well-drained uplands of North Jersey are described. These include a successional group of vegetation and the three forest types – the Mixed Oak forest, the Hemlock–Mixed Hardwoods forest, and the Sugar Maple–Mixed Hardwoods forest type (Figure 10-1). The word "hardwoods" includes deciduous broad-leaved trees which lose all their leaves in winter, as contrasted with the cone-bearing needle-leaved trees, most of which have much softer wood and remain in leaf all winter.

Successional Plant Communities of the
North Jersey Uplands

The important ecological concept of natural change (or succession) in vegetation through time was introduced in Chapter 1 and has been illustrated in previous chapters that deal with plant communities of wetter habitats. Throughout North Jersey, various stages of successional change in vegetation can be seen on the uplands. For the most part these have developed because land formerly cultivated or bulldozed has been abandoned and left untouched. From the time at which the interferences ceased through succeeding years, a series of different plant communities occupy these sites, culminating finally in a mature forest.

An example of the successional changes that take place on the Piedmont uplands of red shale was cited in Chapter 1. A pre-

Mesic Upland Habitats in North Jersey

(Flats, Slopes, Hilltops, Valleys, and Ravines in the Ridge and Valley, Highlands and Piedmont Regions)

One of Three Forest Types Grow on the Undisturbed Mesic Uplands

Community Structure	Mixed Oak	Sugar Maple–Mixed Hardwoods	Hemlock–Mixed Hardwoods	Successional Vegetation — Stages of Succession
Tree Dominants	Red oak White oak Black oak	Sugar maple and many of	Hemlock (Dominant) and only few of	*Annual Herbs* Ragweed Foxtail grass Wild radish Yellow rocket
Other Typical Trees	Chestnut oak Scarlet oak Hickories Red maple Sugar maple Ash Beech Tulip tree	Sweet birch Yellow birch Basswood Beech Ash Red maple Red and white oaks Tulip tree Others	Sweet birch Yellow birch Basswood Beech Ash Red oak Sugar maple Red maple	*Perennial herbs* Aster Goldenrod Little bluestem grass *Initial woody invaders* Red cedar or Gray birch
Tree Understory	Dogwood (Dominant) Sassafras Hop hornbeam Ironwood	Hop hornbeam Dogwood Ironwood Sassafras	Few	Large-toothed aspen and some Wild cherry Sassafras Red maple
Shrubs	Viburnum Spicebush Others	Viburnum Spicebush Others	Few	Shrubs *Young Woodland* Mixed oak or tulip tree stand
Herbs	Many spring & fall herbs	Many spring & fall herbs	Few Partridge berry Mosses	

Man's actions that affect vegetation	Destroy vegetation or modify composition mostly by cutting and fire

Figure 10-1 The Vegetation of the Mesic Upland Habitats in North Jersey.

169

served area near Princeton, New Jersey, also part of the Piedmont uplands, has examples of the successional vegetation that has developed naturally on farmland abandoned seventy to eighty years ago. On this land, part of the Herrontown Woods preserve, the first trees to occupy the abandoned field were red cedar; most of these are now dying off for lack of sunlight. The cedars are being shaded out by taller growing trees such as the sassafras, large-toothed aspen, tulip tree, sweet gum, gray birch, black locust, black cherry, and red maple. All but the last of these trees, the red maple, will succumb in time because all are shade-intolerant and cannot flourish successfully when the trees in the forest become more abundant and grow taller and fuller so that little sunlight penetrates the top canopy.

Farther north in Voorhees State Park, Hunterdon County, in the Highlands section, it has been observed that some fields left idle have immediately been filled with an annual plant called foxtail grass and with ragweed. Whether foxtail grass or ragweed is more abundant in a new field in this area of New Jersey appears to be dependent upon the month in which the field is left idle. Other herbaceous plants include crabgrass, yellow rocket, common mullein, Canada thistle, milfoil, butter and eggs, and a plant called hairy vetch. Also in Voorhees Park a mowed roadside field is filled with low growing herbs of cinque-foil, wild strawberry, dandelion, plantain, a number of grasses (orchard, blue, bent, and crabgrass), and taller more conspicuous flowering herbs—chicory, hawkweed, and goldenrod as well as ragweed.

As on the uplands of the Piedmont, seedlings of young trees invade these fields shortly after they have been abandoned. On other than limestone and particular shale substrates, the red cedar tree is not as typical in the Highlands or the Ridge and Valley sections as it is in the Piedmont abandoned fields. Instead, the pioneer trees in the northern areas primarily are the gray birch and the large-toothed aspen with some wild black cherry, sassafras, and red maple. Along with the first trees, thickets of gray dogwood shrub develop and vines, particularly

poison ivy and Virginia creeper, are as common here as in abandoned fields to the south. Though the species may vary, the sequence of succession is the same on the uplands of the Highlands
as on the Piedmont—first, a growth of annual herbaceous
plants which are replaced by perennial herbs. Then shrubs and
young trees appear spottily throughout the field and in time
these become small thickets interwoven with vines. The thickets
expand and grow taller and what was an open field becomes a
woodland and finally a forest.

In the most northern uplands in the Ridge and Valley section, the successional story is the same though some of the char-
acters differ (Figure 10-2). Land that had been farmed and then
abandoned is invaded first by herbaceous plants, these are replaced with shrub thickets, then woodland grows, and finally the
land is covered with mature upland forests. Examples of all these
stages can be seen at High Point State Park where little bluestem,
Queen Anne's lace, goldenrod, and St. John's-wort are the most
abundant perennial herbs in abandoned fields. The herbs of hawkweed, butter and eggs, and cinquefoil as well as grasses also are
present. The first trees to invade the fields are gray birch and
these gradually form thickets with shrubs of sumac, scrub oak,
blackberry, and various heaths. As time passes, the gray birch
is joined by red maple and wild black cherry trees and more
shrubs—the blueberries, viburnum, witch hazel, blackberry, and
gooseberry. Goldenrod and cinquefoil are common and in places
club moss forms a ground mat. Gradually, seedlings of the oaks
invade the upland area transforming it to a mature forest community of the type described below.

Mixed Oak Forest of the Mesic
North Jersey Uplands

Of the forest cover that now remains on the well-drained
uplands of North Jersey, the most common type is one which can
be appropriately called a Mixed Oak forest. Before the early
1900s, this type of forest actually was known as the Oak-Chest-

Unlike the red cedar succession in the Piedmont (Figure 1-3), in the Ridge and Valley and Highlands regions on other than limestone and particular shale substrates, the first trees to invade abandoned fields usually are the aspen (*large tree at right*) and the gray birch (*2 trees in center*). Both trees can be distinguished by their bark; that of the aspen is olive green and the gray birch has a grayish white bark with black triangular patches below branches. The bark of the latter does not peel off in strips as does that of the paper (or white) birch, a tree not common to New Jersey and found only on northern ridges (Chapter 11). This picture was taken at Voorhees State Park.

Figure 10-2 Successional vegetation of North Jersey.

nut forest for the chestnut trees were as abundant then as the oaks, but now, because of the die-off of the chestnuts, this name no longer is appropriate.

More than fifteen species of oak trees grow in this state but the Mixed Oak forest gets its name from three particular species that in varying mixtures are most abundant among the large trees that form the forest canopy at a height from about 60 to 100 feet. These three are the red oak, the white oak, and the black oak. Two other oaks—the chestnut oak and the scarlet oak—are also present but less common in the North Jersey mesic upland forests than the other three species. Though perhaps initially confusing, the five oak trees can be distinguished rather easily by differences in leaf form, bark, and features of their fruit—the acorns. For those interested, a list of reference material on plant identification is offered in Appendix II.

In addition to the oaks, other large trees that may be present in the mesic upland mixed oak type forest of North Jersey include several types of hickories, red maple, sugar maple, white ash, tulip tree, beech, black cherry, sweet birch, black gum, and elm.

Growing below the tops of the larger trees are smaller trees which may form a distinctive understory layer at a height of about 30 to 40 feet (Figure 10-3). Throughout most of the North Jersey uplands, the dogwood tree is the most abundant of these smaller trees. Also common locally, however, are representatives of the hop hornbeam, sassafras, and ironwood trees, as are chestnut sprouts which grow from the root crowns and are not yet disturbed by the chestnut blight fungus.

Under the two tree layers, the Mixed Oak forest usually has a lower growing shrub layer. In this layer the maple-leaved viburnum which grows to about 3 to 4 feet tall may be very abundant (Figure 10-3), but other common shrubs will include two taller viburnums—the black haw and the arrowwood—the spicebush, witch hazel, the beaked hazel, and red and gray dogwoods; in a few places of very acid soil heath shrubs may grow, including several species of blueberries, huckleberry, and pinxter flower. Poison ivy, Virginia creeper, Japanese honeysuckle, and wild grape are familiar vines of this forest.

The large trees (mostly white, black, and red oak) form a high canopy in this forest and, typically, the flowering dogwood tree forms a distinctive understory layer as shown in the photograph above taken in the Hutcheson Memorial Forest near New Brunswick. The Hutcheson Mixed Oak forest also usually has a distinct lower shrub layer as shown in top photograph on the opposite page. In the foreground is a layer of maple-leaved viburnum shrubs. The spring herb layer in this same forest is dominated by may apple which is shown in the lower photograph on the opposite page.

Figure 10-3 Layering in the Mixed Oak forest of New Jersey.

The herbaceous plants of the Mixed Oak forest vary with time of year as well as with location. In spring the may apple, violets, spring beauties, anemones, jewelweed, jack-in-the-pulpit, solomon's seal, wild sarsaparilla, false lily-of-the-valley, and other early flowering herbs attract attention while in late summer and fall the woodland asters, goldenrod, grasses, sedges, and ferns are more conspicuous. In very moist spots skunk cabbage can be found, and in forest openings where trees are down pokeweed often grows.

Examples of different forms of the Mixed Oak forest can be seen throughout the well-drained uplands of North Jersey (Figures 10-3 and 10-4). In the Piedmont it occurs both on the relatively flat expanses of red sandstone and shale as well as on slopes of the diabase and basalt ridges including the Watchungs, the Sourlands, and the Palisades sill. In the Highlands and the Ridge and Valley regions, the Mixed Oak forest is found on lower slopes and hilltops whether they be made of shale, sandstone, conglomerate, or gneiss. These forests differ from those on the drier ridges and slopes of higher elevation which are described in Chapter 11.

The varying composition of the upland Mixed Oak forest can be illustrated best by descriptions of actual forests in different locations of North Jersey. Starting with the Piedmont region, a representative mature Mixed Oak forest is the Hutcheson Memorial Forest near New Brunswick. The soil here is that derived from the red shale typical of most of the Piedmont. More than 70 percent of the bigger trees in the forest, and these measure as much as 90 feet in height, are oak; in this case, white oak is more abundant than either the black or red oak. Hickory is the fourth most abundant species of the canopy and also present but less common are representatives of sugar maple, red maple, beech, white ash, Norway maple, pin oak, black gum, black cherry, and the lower-growing sweet cherry tree. Dogwood trees form an almost continuous underlayer in this forest at a height of 35 to 40 feet. In better-drained areas the maple-leaved viburnum is the most common shrub, with some black haw present. Spicebush and arrowwood are more abundant in more moist parts of the forest.

Figure 10-4 Two of the more common oak trees of the Mixed Oak forest
in North Jersey, the white oak (*above left*) and the northern red oak
(*above right*), differ in leaf and acorn shape and also the bark of each is
distinctive. In New Jersey a fungus often lives in the inner (dead) part
of the bark of the white oak and weakens the outside, causing peeling
and smooth patches as shown on the right side of the tree above. This
does not appear to harm the white oaks. The bark of the northern red
oak tree as shown is darker than that of the white oak and is broken into
wide, flat-topped ridges separated by shallow fissures. These distinctive
ridges give rise to the description of the red oak bark as having "ski
tracks." The bark of the third most common oak, the black oak, in the
Mixed Oak forest is shown in Figure 12-1.

In areas where winds have blown down trees making openings in the forest, Japanese honeysuckle, poison ivy, and Virginia creeper form thickets. In spring the may apple is the abundant herb on the forest floor but by late summer this role is filled by enchanter's nightshade. Interestingly, it appears that while the oak trees are now the dominant trees in the Hutcheson Forest, there are few young white or black oak trees now growing in the forest. Those that are present are outnumbered by young sugar maple, red maple, Norway maple, and ash trees. The significance of this is discussed later.

In Herrontown Woods, a preserve in Princeton, a mature upland forest on a moist slope is dominated by three oaks—white, black, and red. But in parts of this forest the tulip tree and the sweetgum, the latter more typical of lowlands, also are common. Dogwood is the principal understory tree but ironwood also is common. The maple-leaved viburnum is the most abundant shrub except on moist land where the spicebush is favored. The flowering herbs in the spring include the may apple, spring beauties, jewelweed, jack-in-the-pulpit, solomon's seal, and anemones; in the fall the asters, goldenrod, baneberry, New York fern, Christmas fern, common tick trefoil, and partridge berry are conspicuous.

The forests that remain on the traprocks of the Watchungs and the Palisades in the Piedmont also exhibit the characteristics typical of the Mixed Oak forest type. Throughout the 2,000 acres of the Watchung Reservation the oak forest dominates though its composition varies by site location. On one hillside, red, white, and black oak make up 90 percent of the mature trees with red oak the most abundant tree. With this mixture there are occasional large chestnut oaks and sugar maples. The smaller trees include red maple, white ash, sweet birch, black gum, and dogwood. On another slope, a dry southeast-facing one, white oak and black oak predominate, with tulip tree, red oak, beech, hickory, and maple common but less important. Dogwood is also the most abundant lower tree in this area but with it grow some sassafras and hop hornbeam. Again, the maple-leaved viburnum is the principal shrub. The tallest trees in the Watchung Reservation are found on a northwest-facing slope and while the white, red, and black oaks are still the most abundant trees, the tulip tree is very common. Also scattered through this forest are large numbers of beech, sweet birch, scarlet oak, and red maple. The understory trees include mostly dogwood and sassafras. The shrubs are spotty and consist mostly of the maple-leaved viburnum with occasional black haw, spicebush, and pinxter flower.

The Greenbrook Sanctuary, also on the Piemont but part of the Palisades, has a Mixed Oak forest in which the red oak predominates with white and black oak, sweet birch, sugar maple, tulip, ash, and chestnut oak trees common. Also present in smaller numbers are scarlet oak, elm, red maple, basswood, black cherry, and hickory. The maple-leaved viburnum forms a low, 3- to 4-foot high, shrub layer and there is prolific vine growth in places mostly of poison ivy, wild grape, catbrier, bittersweet, and the moonseed. In this forest, as in others, there appear to be more young sugar maple trees than young oak trees.

On the gneissic and shale slopes of the Highlands the story is the same. For example, in Voorhees State Park of Hunterdon County, the Mixed Oak forest is the most extensive type found. The red oak is the most common tree but is joined by the white, black, chestnut, and scarlet oaks in varying mixtures. Red maple and hickory also are present and on the moister slopes, the beech, sugar maple, tulip tree, ash, and sweet birch trees frequently appear. Dogwoods occur here and form a lower tree layer with hop hornbeam and ironwood. In some locales, particularly on moist gneissic slopes in the area of Morristown, as on the Piedmont traprocks, the tulip tree, a commercially valuable timber tree, grows abundantly. However, this tree appears not to be able to perpetuate itself well in older forests of New Jersey where it must develop under very shady conditions. In this same locale though with a wider range the beech also is a very important tree. In New Jersey the beech perpetuates itself mostly by sprouting from the roots so it tends to persist in an area once it becomes established. In the Highlands, as noted previously for some Piedmont oak forests, the most common younger trees in the upland forest frequently are not oaks but rather consist of red maple, sugar maple, sweet birch, and ash.

In the Ridge and Valley region the forests of ridgetops and

slopes of higher elevations are not included as part of the mesic upland habitat. But the lower slopes and hilltops of this section as well as valleys are considered uplands and here again the Mixed Oak forest is a common forest type primarily on slopes and hilltops that are underlain by shale. Such forests resemble those farther south in that mixtures of the oaks — red, black, white, and here, also, chestnut oak — comprise a majority of the larger trees. Red oak, particularly, appears very abundant in these forests. Also present in larger tree sizes are the red maple, sugar maple, beech, white ash, sweet birch, tulip tree, and black gum. Scarlet oak and hickory are not as abundant but may be important locally. The lower growing trees include the dogwood, hop hornbeam, sassafras, and some striped maple. The shrubs typical of the Mixed Oak forest as far north as High Point State Park include the beaked hazel, witch hazel, maple-leaved viburnum, juneberry, and red and gray dogwoods along with the blueberries and blackberry. In this same area wild sarsaparilla, goldenrod, and ferns make up a rather small herb cover. Again, a striking feature of these Mixed Oak forests is a lack of a well-developed younger generation of oak trees to replace the older trees in the future. Instead, in many stands, red maple, ash, sweet birch, and sugar maple are the more abundant younger trees.

Hemlock–Mixed Hardwoods Forest of the Mesic North Jersey Uplands

Although the Mixed Oak forest now prevails on most of the uplands of North Jersey, a much different appearing forest called the Hemlock–Mixed Hardwoods type occurs on cooler and moister sites located in ravines or on the steep lower, north-facing slopes leading down to ravines or valleys. Sites such as these are found in all three sections of North Jersey — on the Piedmont as part of the traprock formations, in the Highlands on slopes of gneiss, and in the Ridge and Valley section on land underlain by shale, sandstone, or conglomerates.

In all cases the character of the hemlock forest is similar. Typically, more than half the larger trees of the forest are hemlocks. The hemlock grows more abundantly to the north of New Jersey in New England and New York State. The small needles of the hemlock are dark green and the tree is evergreen, retaining the needles throughout winter. Of the other large trees associated with hemlock, some also are more typical of northern than of southern forests. These include the sweet birch, yellow birch, sugar maple, and basswood. But also common in the hemlock forest are some of the same trees that are found in the Mixed Oak forest, typically the beech, red oak, white ash, and red maple. The fallen needles of the hemlock create a very acid condition on the forest floor, which, combined with the lack of sunlight year round, appears to discourage the development of dense undergrowth cover, whether of lower trees, shrubs, or herbs (Figure 10-5).

In the Watchung Reservation in the Piedmont region, for example, a hemlock-dominated forest occurs on both a northwest-facing valley slope and in a rocky ravine. In both, hemlock accounts for more than half of the trees in the forest with scattered representatives of beech, sweet birch, red maple, sugar maple, white oak, red oak, and black oak. In the Hemlock–Mixed Hardwoods forest, unlike the Mixed Oak forest, dogwood occurs only infrequently as an understory tree. The shrub cover is not dense and herbs are almost absent. A hemlock forest of about the same mixture of species also occurs on ravine slopes in the Greenbrook Sanctuary on the Palisades. And again there is no dogwood layer. The shrubs, mostly the maple-leaved viburnum, spicebush, and blueberry, are sparse and scattered. There are some vines — poison ivy, Virginia creeper, and the wild grape — and a few herbs covering the ground, particularly the partridge berry, false lily-of-the-valley, and mosses.

The hemlock forests of the lower north-facing valley slopes and ravines of the Highlands and the Ridge and Valley regions are much the same as those described for the Piedmont. The hemlock tree not only dominates the canopy but it also successfully reproduces and develops as a smaller tree in the forest. Red oak, sweet birch, chestnut oak, black oak, red maple, beech, and yellow birch are the most consistent associates of the hemlock with white oak, sugar maple, black gum, tulip tree, and ash appearing only

Top picture taken in Tillman Ravine at Stokes State Forest typifies the site and character of the hemlock forest—a ravine having mostly hemlock trees and little undergrowth except for the rhododendron (*at left*). In contrast, the Sugar Maple-Mixed Hardwoods forest, such as the one shown in the picture (*bottom*) of a forest in a valley at High Point State Park, is a very rich forest with many different tree species (*sugar maple is large tree at right*) and a varied shrub and herb layer. The sugar maple is reproducing itself well as shown by the many seedlings in the forest. Leaves at the base of the tree trunk at the right are those of a maple seedling.

Figure 10-5   Hemlock-Mixed Hardwoods and Sugar-Maple Mixed Hardwood forest types

as scattered members of the forest. In none of these forests is there a well-developed understory of trees other than the young trees of hemlock, birch, and red maple. Striped maple and hop hornbeam appear only occasionally and dogwood seldom. The shrub layer also is poorly developed with witch hazel and maple-leaved viburnum most common and locally some rhododendron. The herbs are sparse and for the most part repeat those already mentioned.

Sugar Maple–Mixed Hardwoods Forest of the Mesic North Jersey Uplands

It is unfortunate that most of the fertile limestone valleys in North Jersey have been cleared of natural vegetation. For it is on such sites that the Sugar Maple–Mixed Hardwoods forest flourishes. When this type of forest occurs, it is diverse, with representatives of many tree species and with a lush undergrowth. Examples of this forest are found mostly on the Kittatinny limestone underlying the Great Valley in the Ridge and Valley region. Sugar maple is the most abundant tree in this type of forest, but associated with it in large numbers are hardwood trees typical of areas to the north and also some of those more typical of the Mixed Oak forest. Altogether the Sugar Maple–Mixed Hardwoods forest is more diverse than the Mixed Oak or Hemlock–Mixed Hardwoods forest—it is richer in numbers of plant species and its membership is more equitably distributed among species.

White, black, and red oak are common in the Sugar Maple–Mixed Hardwoods forest, as are white ash, tulip tree, sweet birch, yellow birch, red maple, basswood, beech, and the hickories. Scattered members of hemlock, white pine, elm, black walnut, and another oak—the yellow oak—can be found in some of the wooded areas. Interestingly, however, hemlock and sugar maple do not occur together with great abundance of each. Where hemlock is dominant, sugar maple is sparse and the reverse also is true. Unlike hemlock-dominated forests, the sugar maple–dominated forests normally exhibit a well-developed layer of lower growing trees (Figure 10-5). Although some dogwood is present in these forests growing even in northernmost New Jersey, hop hornbeam appears to be more common. Sassafras and ironwood trees also are present in this layer. Shrubs are abundant and include the maple-leaved viburnum, black haw, spicebush, and beaked hazel. Many early spring flowering herbs grow on the floor of this forest, and unusual ferns can be found in crevices of the limestone outcroppings.

One of the forests in High Point State Park illustrates well the composition of the Sugar Maple–Mixed Hardwoods type community. The trees are tall and straight, reaching heights of over 90 feet. Sugar maple is the most abundant tree, but some of its associates also are common, including beech, yellow birch, sweet birch, red oak, white oak, and basswood. That chestnut also was important once in this forest is indicated by huge logs of this tree that rest on the forest floor. An understory tree layer consists mostly of hop hornbeam and dogwood trees with young sugar maples that are reproducing abundantly.

While the shrub layer is not dense, there is more coverage than in the hemlock-dominated forest. Spicebush is the most common shrub, and others in decreasing importance are the witch hazel, maple-leaved viburnum, and beaked hazel. Poison ivy and Virginia creeper vines also are present. Jack-in-the-pulpits are conspicuous in early spring and skunk cabbage in wet areas. These with some ferns, mosses, and other herbs, compose the ground layer.

The same plants more or less occur together in other forests in or adjacent to the Great Valley. In some, the tulip tree may be as common as some of those already mentioned. In all the Sugar Maple–Mixed Hardwoods forests just mentioned, the sugar maple tree is reproducing prolifically and, as a result, is abundant among the smaller trees. The same is true of the other northern species — the birches, basswood, and hemlock. Except for red oak, the oaks in these forests appear less successful in assuring themselves a place in the next generation of canopy trees.

**Upland Vegetation of North Jersey
without Man's Influence**

Ecologists and foresters have speculated about the probable composition of the upland vegetation in North Jersey if man had not settled here. Obviously there would be a much larger amount of natural vegetation than there is. But the present vegetation might be less varied if man had never settled in the state. One reason this is so is that all the uplands would be completely covered with mature forest, and the successional vegetational stages stemming from the abandoned fields would be absent.

Another question arises with respect to the nature of the forest composition if the land had not been disturbed by fire or by cutting. Some ecologists believe that under present climatic conditions the prevailing vegetation on the uplands would be Sugar Maple–Mixed Hardwoods forest except on ridges or on those particular sites especially favorable for the development of the Hemlock–Mixed Hardwoods forest. This conclusion is based on evidence that when fire and excessive cutting of the woodlands are controlled, species of sugar maple, yellow birch, sweet birch, and hemlock appear to be able to invade successfully forests which previously were dominated by white, red, or black oak trees. Certainly, if present abundance of young tree seedlings and saplings indicate the trees that will dominate the future composition of the forest, this conclusion would appear to be correct. Thus, if man had never settled in New Jersey, instead of the Mixed Oak forest being the most common type on the North Jersey uplands, the Sugar Maple–Mixed Hardwoods forest type would be more abundant. In this sense, then, the present Mixed Oak forest may represent a successional stage of forest which, if left undisturbed, eventually and naturally will progress into one of the other two forest types.

Some authorities question the validity of this speculation with respect to the dominance of sugar maple on all mesic uplands in North Jersey. It may be that the large numbers of young sugar maple trees do not necessarily reflect the future composition of the forest. And although the hemlock and sugar maple and the associated northern hardwoods may be more successful than the oaks on cooler and moist sites in northernmost New Jersey, these trees may never do as well as the white, black, and red oak under warmer and drier conditions, as on Piedmont red shale sites or on south-facing slopes in the Highlands and the Ridge and Valley sections.

Such speculation is interesting but can never be resolved. Only if the present forest lands were maintained without disturbance for hundreds of years would it be possible to determine what New Jersey forests would be like without man.

Summary

The topography and the soil parent rock material of the North Jersey uplands varies widely. But the diversity found in the natural vegetation stems primarily from the actions of man. The vegetation of four types of plant communities is described in this chapter – successional stages of vegetation, the Mixed Oak forest, the Hemlock–Mixed Hardwoods forest, and the Sugar Maple–Mixed Hardwoods forest. It is speculated that only the last two types of vegetation would still exist if men had never settled in the area.

REFERENCES AND SOURCE MATERIAL

Baird, J. 1956. The Ecology of the Watchung Reservation, Union County, New Jersey. Rutgers University, New Brunswick, N.J.

Buell, M. F., Langford, A. N., Davidson, D. W. and L. F. Ohmann. 1966. The Upland Forest Continuum in Northern New Jersey. Ecology 47: 406–431.

Collins, Stephen. 1956. The Biotic Communities of Greenbrook Sanctuary. Ph.D. Thesis, Rutgers University, New Brunswick, N.J.

Kramer, Richard J. 1971. Herrontown Woods. Stony Brook-Millstone Watersheds Association, Inc. Pennington, N.J.

McDonough, W. T. and M. F. Buell. 1956. The Vegetation of Voorhees State Park, New Jersey. American Midland Naturalist 56: 473–490.

Monk, C. D. 1961. The Vegetation of the William L. Hutcheson Memorial Forest, New Jersey. Bulletin of the Torrey Club 88: 156–166.

Niering, W. A. 1953. The Past and Present Vegetation of High Point State Park, New Jersey. Ecological Monographs 23: 127–148.

Pearson, P. R. 1960. 1961. Upland Forests on the Kittatinny Limestone and Franklin Marble of Northern New Jersey. Bulletin of the New Jersey Academy of Science 5: 3–19.

Vermeule, C. C., A. Hollick, J. B. Smith and G. Pinchot. 1900. Report on Forests in Annual Report of the State Geologist for 1899, Trenton, N.J.

11

Vegetation of the North Jersey Ridgetops, Slopes, and Rock Outcroppings of Higher Elevations

Introduction

Excluded from the category of well-drained uplands discussed in the previous chapter are two drier habitats — ridgetops and slopes of higher elevations and rock outcroppings. The ridgetops and slopes of higher elevations are drier than mesic upland sites for two reasons. First, rainwater is quickly lost because of rapid runoff, and that which does penetrate the soil may evaporate quickly because of exposure to sun and to wind. In addition, the soil cover on these sites is usually thin, infertile, and of poor water-holding capacity. But still drier than the ridgetops and slopes are the bare rock outcroppings which without a mantle of soil make a more rigorous type of plant habitat to which only a few species appear well adapted.

This chapter is divided into two sections; the first is devoted to the plant communities of the ridgetops and slopes of higher elevation in North Jersey (Vegetation Type 8) and the second covers the vegetation that grows on rock outcroppings (Vegetation Type 9).

188

Vegetation of the Ridgetops and Slopes of Higher Elevations in North Jersey (Vegetation Type 8)

The vegetation characteristic of ridgetops and slopes of higher elevations in New Jersey is found mainly in the Highlands and the Ridge and Valley regions though the ridge formations of diabase or basalt on the Piedmont such as the Watchungs have examples of the same plant growth.

Two fairly different plant communities can be delineated on these sites, though gradations between the two also occur. Both communities are forest types; the first, referred to as the Chestnut Oak forest, is the vegetation type that prevails on most of the slopes and ridgetops of higher elevations in North Jersey. The second community, the Pitch Pine–Scrub Oak forest, is characteristic only of the highest ridgetops in the state and is found on Kittatinny Mountain.

Chestnut Oak Forest

Although the chestnut oak tree grows in the Mixed Oak upland forest described in the previous chapter, it is not one of the more abundant trees in this type of forest. But on slopes at higher elevations the chestnut oak becomes the most important tree in the plant community. It appears to be able to reproduce and develop better than the other oaks under the drier and poorer soil conditions characteristic of the higher slopes. In appearance the Chestnut Oak forest is quite different from the Mixed Oak forest described in the previous chapter. The trees are not as tall, growing only to a height of about 50 feet, and the crowns of the trees do not form a completely closed canopy. Because of this, considerable sunlight penetrates to lower levels in the forest. Where a slope is rocky, as many are in northern New Jersey, trees grow in a stunted fashion, often with their tops broken off by ice storms (Figure 11-1).

The second most abundant tree in the Chestnut Oak forest may be any one of five trees—the red oak, the white oak, the

Picture above shows the growth form of the Chestnut Oak forest on top of Kittatinny ridge where the Appalachian Trail crosses the ridge. The tops of the trees are broken and stunted by the severe wind and ice storms. Picture on opposite page shows the forest as it appears in a protected area on Kittatinny ridge in High Point State Park. Typically, the chestnut oak tree (*right foreground*) has dark bark that is very deeply and coarsely furrowed. The foliage to the left of the chestnut oak is that of a chestnut tree coming from sprouts. Blueberry shrubs are in the foreground.

Figure 11-1   Chestnut Oak forest type.

191

scarlet oak, the sweet birch, or the pitch pine. The last three trees are more typical of the sites with the thinnest soil cover. In addition to these species, the black oak, red maple, hickory, black cherry, and white pine in smaller numbers are frequently associated with the chestnut oak. Evidence that the American chestnut also was once part of this group is seen by the sprouts which continue to come from the diseased tree trunks. Some dogwood and sassafras are found as lower growing trees in the community but do not form a continuous understory as in the Mixed Oak forest.

The shrubs of the Chestnut Oak forest most typically are the heaths, plants typical of acid soils; these include the blueberries, huckleberry, and laurel. There are but few herbs on the forest floor.

Many forms of the Chestnut Oak forest can be seen on the higher slopes in areas of the Ridge and Valley and the Highlands regions underlain with conglomerate or sandstone parent rock material (Figure 11-1). One such slope is in High Point State Park where the chestnut oak is observed to contribute about 55 percent of the tree cover. The larger trees measure 9 to 14 inches in diameter, average about 50 feet in height, and are approximately 95 years old. On this particular site, red oak is the second most abundant tree with black oak and sweet birch also common. The understory tree layer is composed of red maple, sassafras, chestnut sprouts, birch, hickory, and many young chestnut oak trees. Shrubs are rather uniformly and continuously distributed in this forest with different plants important in local areas. In one spot huckleberry and blueberry may be most important, and in another mountain laurel and the pinxter flower are more abundant. The herbs are very sparse with wild sarsaparilla and wintergreen the most common.

The highest slopes of the Piedmont such as those of Cushetunk Mountain do not approach the elevation of those in the Highlands or the Ridge and Valley sections but still a form of the Chestnut Oak forest can be found on the ridgetop and higher slope areas. An ecologist studying the vegetation on the upper slopes of Cushetunk, a ridge which is about 600 feet higher than the surrounding Piedmont land, observed differences in the climate and in the vegetation between the north-facing and the south-facing slopes of the mountain. The air temperature, particularly close to the ground, is consistently higher on the south side, which receives a greater amount of solar radiation. The climatic differences between the two slopes are reflected in the vegetation in several ways. For example, more large-sized trees grow on the north slope than on the south. And although chestnut oak and red oak are the principal trees on both slopes, sweet birch, tulip tree, ash, basswood, and sugar maple trees are more abundant on the north side than on the south. In contrast, black oak, white oak, and hickory are more important on the south-facing slope than on the north. Among the lower growing trees, dogwood is much more important on the south slope than on the north, but twice as many shrubs grow on the north side which also has more variety of species in this layer. The greatest difference in species composition between the north-facing and south-facing slopes actually occurs in the low growing herbs. Asters, goldenrod, and hog peanut are the most abundant herbs on both slopes, but the slopes differ greatly in the composition of the remaining herbs. On the north slope, plants more typical of northern forests grow — wild ginger, wild sarsaparilla, black snakeroot, columbine, ferns, and mosses. On the warmer south slope, various species of grasses, sedges, and annuals are more abundant.

Pitch Pine–Scrub Oak Forest

On the highest ridgetop in New Jersey, Kittatinny Mountain, can be seen a community in which the pitch pine is the most abundant tree. It grows here on a very infertile soil high in silica which is similar in development to the soil of the Pine Barrens — the characteristic location of the pitch pine in New Jersey. On the ridgetop the soil cover not only is poor and thin but climatic conditions are rigorous. The site is exposed to frequent sleet and ice storms and strong winds. Similar conditions prevail only on a few ridgetops in the state; the state geologist in 1899 reported clumps of pitch pine growth on fire-damaged ridges of the Bear-fort, Green Pond, and Pohatcong mountains.

Several forms of the pitch pine community can be found in High Point State Park (Figure 11-2). Under the most rugged conditions, the pitch pine is the only abundant tree and it accounts for as much as three-quarters of the total number of trees. The other trees which occur in sparse numbers, include the red maple, sweet birch, gray birch, and a low growing tree or shrub called juneberry. The result is hardly a forest in appearance for the pitch pine trees average only about 20 feet in height even though they are 70 years old, and are sparsely distributed. The result is a forest of very open canopy. Because of the ice, sleet, and wind storms many of the treetops are broken off and many branches twisted. Trees that take root in rock crevices grow at various angles to the slope. In the lower stratum of the forest, the ground is mostly covered by thickets of the scrub oak which reach no more than 10 feet high. Lower-growing shrubs, mostly blueberry and black huckleberry with some sweetfern, laurel, sumac, and black chokeberry, also are present. Herbs cover about 20 percent of the ground in this forest with wild sarsaparilla, bracken fern, and false lily-of-the-valley most common.

On more protected parts of Kittatinny Mountain, the Pitch Pine–Scrub Oak forest is better developed. In these, the pine remains the most abundant and tallest tree, but chestnut oak, scarlet oak, and white oak trees in varying mixtures are common along with the sweet birch and red maple. Occasionally, specimens of black gum, black oak, and sassafras also occur. Seedlings of oak trees often are more common than pine, which leads to the speculation that with control of fire the oaks may some day dominate the forest. The reason for this is that the pitch pine appears more resistant to fire damage than the oaks.

Scrub oak is not as common on the more protected ridgetop sites where huckleberry is the dominant shrub. Associated with it are the same heath shrubs found on the less protected ridgetop. The herbs are sparse in number and only the bracken fern is abundant.

Vegetation of the Rock Outcroppings of North Jersey (Vegetation Type 9)

On the soil moisture scale the habitat of rock outcroppings of North Jersey is equated with that of the sand dunes of South Jersey. Both represent the extreme of dry habitat conditions in the state.

The higher ridgetops in northern New Jersey were stripped of their soil mantle and vegetation by the glacial ice sheets, leaving exposed the bare rock surfaces. Since the disappearance of the ice sheets thousands of years ago, some parts of the exposed rocks have become covered with a soil mantle and with plant growth. The process by which this has happened represents another example of plant succession. The rate at which the surface of bare rock is covered by vegetation is extremely slow, though it varies with climatic and substrate conditions. For this reason, some parts of the ridgetops in North Jersey are still exposed rock outcroppings, others are in various stages of vegetation succession, and still others already have a covering of Chestnut Oak or the Pitch Pine–Scrub Oak forest. For purposes of describing the vegetation on the rock outcroppings in North Jersey, four idealized stages of successional plant communities are identified, but in actual life gradations and overlapping of the stages occur on the same rock. The four stages are: (1) the lichen and moss invasion, (2) the herbaceous plant invasion, (3) the shrub invasion, and (4) the initial establishment of tree invaders. Examples of all these can be seen within a short distance on rock outcroppings in High Point State Park (Figures 11-3, 11-4 and 11-5).

Lichen and Moss Invasion of Rock Outcroppings

Few plants are able to survive the environmental conditions found on the exposed rock surfaces on the ridgetops of New Jersey. Without the moderating influence of a plant and soil cover, extremes of temperature occur on bare rocks. During the

On the ridgetop (*above*) the pitch pine trees are sparse and windshapen. Scrub oak and heath shrubs grow under the pine trees. Picture taken in High Point State Park overlooking Lake Marcia. On *opposite page*, pitch pine trees growing in a more protected area on the ridge.

Figure 11-2   Pitch Pine-Scrub Oak Forest on top of Kittatinny Mountain.

197

daytime, because of direct exposure to sun, a rock surface becomes very warm but then it quickly cools off at night. Little water and nutrients are available because of the lack of a soil mantle. Only lichens can survive under such conditions, and these are barely recognizable as plants.

The first plant life to invade exposed rock outcroppings are the hardy crustose lichens which appear as dark gray or black stains on the rock surface (Figure 11-3). It is hard to believe that these stains actually represent a partnership of two living organisms, but a lichen consists of a fungus which envelops an alga. The alga contains chlorophyll and can manufacture food for both organisms. The fungus appears to contribute its part by providing an anchorage and a source for absorbing moisture. On the exposed rock outcroppings of the conglomerate rock around the High Point monument in North Jersey there is one abundant crustose lichen with the scientific name of *Rinodina,* and another is commonly known as brownie's buttons. The early lichen settlers aid in the decomposition of the rock material which is transformed into soil. Also, the dead remains of these lichens provide organic material for the next plants to come. There are other types of lichens — the foliose lichens inhabit a bare rock surface.

One species of foliose lichen commonly found on the ridge outcroppings in northern New Jersey is called the rock tripe. One of these, the rough rock tripe, is a brown and leathery lichen most of the time but turns green after a rain. Unappetizing as they may appear, it is said that the rock tripe lichens were used as food at time of famines by Indians and early European settlers. Other foliose lichens found in North Jersey include the smooth rock tripe and the common rock lichen. In some places the reindeer moss, a fruticose lichen that is very abundant in the tundra vegetation of the arctic, can be found. The lichens aid in the accumulation of soil particles on the rock outcroppings; also as old plants die, their remains are incorporated into the soil, enhancing its fertility.

But happy pioneers as they are, the lichen invaders make an insignificant contribution to the invasion of rock surfaces by

Figure 11-3 Plant succession on rock outcroppings. A foliose lichen appears as a stain on a rock at High Point State Park. *Photograph by William A. Niering.*

vegetation. It is the mosses with their bulkier life-form extending as thick mats over rock surfaces that contribute most to initial succession on rocks. Dust and debris collect in these mats. In addition, through the years leaves and old stems of mosses accumulate and this organic material combined with the dust and debris starts the build-up of soil in which herbs and seedlings of shrubs and trees can become established. Mosses start to grow on the rocks mostly as small tufts in crevices where some soil has accumulated. Gradually, they extend outward to form mats over the rock surfaces. The pin cushion mosses, haircap mosses, pigeon wheat moss, thread moss, and rock moss are just a few of the species that can be seen on the Kittatinny ridge.

Herbaceous Plants Invade the Rock Outcroppings

As more windblown dust particles are trapped by the stems of mosses and as the original moss invaders die, the thickness of the moss mat continues to expand over the rock surface. Also, the presence of more plants hastens the disintegration process of surface rock and of the trapped particles of debris. These, combined with the dead plant remains, form the beginning of a soil mantle.

When enough soil has accumulated in any one place to provide both an anchorage for plant roots and some water-holding capacity, herbaceous plants will start to develop successfully from seeds that blow onto the rocks. One of the first herbs abundant on the rock outcroppings in northern New Jersey is called hairgrass (Figure 11-4). The three-toothed cinquefoil, a smaller herb with a tiny white flower, grows in protected crevices. Beardgrass, poverty grass, impoverished grass, sedges, wild sarsaparilla, the rock polypody fern, and the marginal shield fern are other herbs that grow on the moss lichen mat or in rock crevices.

Shrub Invasion of Bare Rock Succession

As herbaceous plants start to occupy the moss-lichen mats on rock outcroppings, a few low growing shrubs, and even trees, will start to grow on the mat or in pockets of soil in crevices. These have their origin from seeds or from the spread of underground systems from plants growing in better soil pockets adjacent to the rocks. The lower growing species of blueberry, black huckleberry, sheep laurel, chokeberry, juneberry, and sumac are among the more common shrub pioneers.

As more shrubs occupy the rock outcroppings, they shade the lower herbaceous plants which cannot survive successfully without full sunlight. Thus, the vegetation changes in its composition and as it does, more and more organic material accumulates, creating more fertile soil conditions. The scrub oak, the typical shrub cover in the forest of ridgetops, grows successfully

Figure 11-4 Herbaceous plants such as hairgrass invade the rock out-croppings. The hairgrass (*center*) grows abundantly where soil has accumulated in the crevices. Small patches of moss can be seen in the crevice at the right. Shrubs are invading the rock from soil pockets at the side of the outcropping. Picture taken near the High Point Monument which appears in the background.

only when the mat of vegetation covering the rocks has become relatively thick.

Shrubs, by virtue of their greater bulk of annually produced litter and their many perennial woody stems that help hold the new "soil" in place, provide an increasing substratum for other plants to grow in. Thus the vegetation successional process gradually changes the habitat; what started as a bare rock surface is transformed into a soil-covered rock substrate on which more kinds of plants can grow. By the very nature of this change, the original pioneer plants which flourish best in the initial stages of rock succession give way to the higher growing shrubs and trees.

Figure 11-5 White birch trees and the rhododendron shrub have become established on rock outcroppings at High Point State Park. *Photograph by William A. Niering.*

Tree Invasion of the Rock Outcroppings

Gradually the soil moisture and fertility conditions improve to the point at which tree seeds that fall on the mat or in crevices will successfully germinate. When the seed of a tree such as the sweet birch tree germinates on a young moss mat, the roots of the tree seedling will grow out through the mat. If they are able to reach a crevice with some soil, the seedling has a good chance to develop into a tree (Figure 11-5).

Among the first trees to grow in the rocky habitat of North Jersey are members of the Pitch Pine–Scrub Oak or the Chestnut Oak forest, especially the pitch pine, chestnut oak, and sweet birch trees, though some representatives of the more northern paper birch tree also can be found growing among the boulders at

202

Drier Habitats of North Jersey

Community Structure	Plants of the Steep Slopes & Ridges		Plants Growing on Rocks
	Chestnut Oak Forest	Pitch Pine–Scrub Oak Forest	Successional Stage of Vegetation
Common Trees	Chestnut oak (Dominant) Red oak White oak Scarlet oak Sweet birch Pitch pine	Pitch pine (Dominant)	*Lichen Moss Invasion* Crustose Lichens (*Rinodina*) Foliose Lichens (Rock Tripes) Mosses (Pin Cushion) *Herb Invasion* Hair grass Cinquefoil Sedges & grasses Ferns *Shrub Invasion* Blueberry Huckleberry Laurel *Tree Invasion* Pitch pine Chestnut oak Sweet birch White birch
Other Typical Trees	Black oak Red maple Hickory Black cherry White pine	Red oak Sweet birch Gray birch Chestnut oak White & scarlet oaks	
Understory	Chestnut sprouts Laurel Blueberry	Scrub oak (Dominant) Blueberry Huckleberry	
Herbs	Few Wintergreen Wild sarsaparilla	Few Bracken fern Wild sarsaparilla	
Man's actions that affect vegetation	Destroy vegetation or modify composition mostly by cutting and fire		

Figure 11-6   The Vegetation of the Drier Habitats in Northern New Jersey.

High Point State Park. For the most part, the pioneer trees are stunted and crooked in shape, and some of their roots may completely straddle large boulders or run almost horizontally in the mat of vegetation covering the rock surface. Tree growth causes greater accumulation of soil on the mat but more shading of the ground layer, and so, as higher plants grow on the site, the original herbs disappear. Better developed trees follow the early invaders and inevitably the vegetation evolves to some form of the ridgetop Pitch Pine–Scrub Oak forest or Chestnut Oak forest.

Summary

The plant communities of two habitats have been described in this chapter (Figure 11-6). The first, that of the ridgetops and slopes of higher elevations, represents a drier habitat than that of the uplands. Its vegetation typically is represented by one of two forest types – the Chestnut Oak forest which prevails over most of the ridgetops and slopes of higher elevations in North Jersey and the Pitch Pine–Scrub Oak forest which grows only on the highest elevations in New Jersey. A still drier habitat is that of rock outcroppings which owe their origin to the scouring action of glacial ice sheets. But even these habitats through the process of succession become inhabited by vegetation with a resulting change in the site conditions. Passing through gradations of four successional plant communities – the lichen-moss stage, the herb stage, the shrub invasion, and the initial tree pioneers – the habitat conditions are so transformed that ultimately the ridgetop forest type of vegetation develops on what was once exposed bedrock.

REFERENCES AND SOURCE MATERIAL

Buell, M. F., A. N. Langford, D. W. Davidson and L. F. Ohmann. 1966. The Upland Forest Continuum in Northern New Jersey. Ecology 47: 406–431.

Cantlon, J. E. 1953. Vegetation and Microclimates on North and South Slopes of Cushetunk Mountain, New Jersey. Ecological Monographs 23: 241–270.

Niering, W. A. 1953. The Past and Present Vegetation of High Point State Park, New Jersey. Ecological Monographs 23: 127–148.

Vermeule, C. C., A. Hollick, J. B. Smith, and G. Pinchot. 1900. Report on Forests in Annual Report of the State Geologist for 1899, Trenton, N.J.

12

Vegetation of the South Jersey Coastal Plains – Its Mesic Uplands and Drier Pine Barrens

Introduction

An arbitrary division of the state into North and South Jersey for the purposes of this book places the boundary coincident with the dividing line that separates the Piedmont from the Inner Coastal Plain (Figure 6-2). The justification for the division between the two areas is supported by significant differences in geologic and soil features as well as in climatic conditions as described in Chapter 6.

A floristic division between North and South Jersey based on the boundary line between the Piedmont and Coastal Plain regions has long been recognized by botanists. In 1910 Dr. Witmer Stone wrote that the boundary "marks a great change in plant life." He identified 1,373 species of plants growing on the Inner and Outer Coastal Plains. Of the total, only 91 species were thought to be widely distributed over both the Coastal Plain and the Piedmont regions. Another 807 species more or less common to the Piedmont region were found only in the area of the Inner Coastal Plain and only 181 of these reach the Pine Barrens. The final 475 species were said to be restricted to the Coastal Plain region except for sporadic occurrences in the Piedmont and then mostly on lowlands. That many species of northern affinity reach the southern limit of their range on the Coastal Plains and are joined there by species of southern affinities reaching their northern limit has made this area of New Jersey an unusual natural laboratory for plant scientists.

The topography of the Coastal Plain does not have the extremes in relief found in North Jersey, for the land varies only from sea level to a maximum elevation of less than 400 feet. Nor is the land underlain by the variety of hard rock material found in the northern part of the state. Instead, the South Jersey Coastal Plains consist mostly of sand, silt, gravel, and clays that are not cemented together in rock form except on the tops of the cuesta hills. There is also a vast difference in geologic age between North and South Jersey; South Jersey is considerably younger, as part of the Coastal Plains was ocean floor until 1 million years ago.

On both the Inner and Outer Coastal Plains there is a considerable amount of lowland, primarily because of the low relief and high water tables. The plant communities occupying these wetter lands—the marshes, bogs, swamps—are described in previous chapters. In addition to these lowland habitats there are sites in South Jersey that can be classified as well-drained uplands. Just as in North Jersey, these mesic sites represent an idealized midpoint of soil moisture—a site with no standing water but with ample moisture in the soil for plant growth. Thus, the mesic uplands in South Jersey mark more or less a halfway mark between the wettest habitat, the marsh, and the extremely dry sand dunes. Unlike North Jersey, however, only a small proportion of the total land of the Coastal Plains can be classified as mesic uplands. This is true for several reasons. First, as mentioned previously, extensive areas of lowlands exist in the southern part of the state. And, second, much of the remaining land is so dry that it is considered xeric rather than mesic.

As described in Chapter 2, geologically the Coastal Plains are divided into two sections—the Inner and Outer Coastal Plains—based on the geologic time at which the sand, gravel, silt, and clay were originally deposited by the seas that covered the land.

As shown in Figure 2-4 the area designated as the Inner Coastal Plain is considerably smaller than the Outer Coastal Plain though together they account for more than half of all New Jersey.

Though the division of the Coastal Plain into the two sections—Inner and Outer—appears clear-cut, based on the age of the underlying strata of clay, silt, sand, and gravel, complications arise from the fact that both sections were later covered by interglacial deposits. The result is that the present soils of the Coastal Plains are varied both in water-holding capacity and in fertility. These two soil characteristics cause the differentiation between the mesic upland and the drier upland habitat in South Jersey.

The mesic uplands include primarily the area of the Inner Coastal Plain but extend also into the most southern, eastern, and northern sections of the Outer Coastal Plain, thereby encircling the drier habitat commonly known as the Pine Barrens (Figure 2-4). Thus, the famous Pine Barrens of New Jersey represent an island of drier and less fertile sandy soil surrounded by more moist and fertile uplands. The dryness of the Pine Barrens results from its coarse, sandy soil through which rainwater readily passes with little retained by the soil. In contrast, the finer-textured soils of the Inner Coastal Plain retain adequate amounts of water creating a mesic substrate for plants.

This chapter starts with a description of the vegetation of the mesic uplands of South Jersey (Vegetation Type 10) and is followed by a look at the more xeric plant habitats in the fascinating Pine Barrens (Vegetation Type 11). The vegetation of a still drier habitat in South Jersey, that of the sand dunes, is covered in Chapter 13.

Vegetation of the Mesic Uplands of South Jersey (Vegetation Type 10)

The area designated as the South Jersey mesic uplands generally has soil more fertile, more moist, and less sandy than that of the Pine Barrens, but a wide variety of soil types occurs within both categories.

The most fertile of the mesic upland sites and those excellent for farming include most of the land making up the Inner Coastal Plain. It is overlain by a north-south oriented belt of soils containing deposits of the mineral glauconite — greensand marl as it is commonly called. In some places the mineral is so abundant that the soil is a dark green color; in other areas it may give simply a green tinge to the earth; and where it is sparsely distributed, the soil is its normal color. Outside the greensand marl area the mesic uplands soil may have more or less clay, silt, sand, or gravel in a variety of different combinations. But as is true in the case of the North Jersey uplands, the same groups of plants appear to grow naturally on a wide spectrum of soil types provided the ground moisture conditions and fertility are about the same. However, it is still difficult to pigeonhole the natural vegetation of the South Jersey uplands since little natural growth remains on the land.

It is the Inner Coastal Plain of the South Jersey uplands that serves as the New Jersey transportation corridor for megalopolis traffic. This is the location of the north-south turnpike and major railroad beds. The land of this area that is not used for transportation facilities or for industrial or residential buildings has been cleared and is devoted to fruit, vegetable, dairy, or poultry farming. For the most part, it is only the wetter lowlands that contain extensive tracts of natural vegetation. Even descriptions of South Jersey upland vegetation made earlier in this century comment on the exploitation of the natural vegetation on the Inner Coastal Plain. As early as 1900 this area was reported to be only 15 percent wooded and the most recent map of forest distribution in the state shows little wooded acreage remaining on the Inner Coastal Plain (Chapter 4, Figure 4-5).

The fragments of natural vegetation that exist on the mesic uplands in South Jersey today reflect the influence of man's actions as does its counterpart in North Jersey. Virgin forest has disappeared and the woodlands that remain have been cut over repeatedly since European settlement of the state. Basically, the present well-drained upland forests in South Jersey are dominated by oak trees with more or less the same composition of that described as the Mixed Oak forest type of the North Jersey uplands. In addition to this forest growth, there are various stages of successional vegetation on land that had been cultivated or otherwise disturbed but then left abandoned.

Mixed Oak Forest of the Mesic South Jersey Uplands

At the close of the last century the state geologist reported that the mesic upland forest of the Coastal Plains was a mixed one, "mostly oak and chestnut"; one such forest on the Navesink Highlands in the northeastern part of the Coastal Plain was reported to have as its principal trees the chestnut tree and the five species of oak that are commonly found in the North Jersey upland forest—white oak, black oak, red oak, chestnut oak, and scarlet oak. Unlike the typical upland oak forest of North Jersey, however, the Navesink forest has a shrub cover of heath plants primarily the laurel, blueberries, huckleberry, and swamp azalea. (See Appendix III, Part A, for listing of New Jersey heath plants.) All these are plants that are typical of sandier or more acid soils. Just as in North Jersey, mature chestnut trees have disappeared from the southern upland forests, though sprouts still come forth from some of the larger stumps.

The forest composition just described for the Navesink Highlands is typical of sandier areas of the South Jersey mesic uplands that are adjacent to the Pine Barrens. The upland forests located on the more fertile soil in the western and southwestern part of the Coastal Plains show a variation in composition particularly in the area from western Monmouth County southward through western Burlington and Camden counties. Here although the oaks, particularly the white, red, and black oaks, are still the most abundant larger trees, the smooth gray-barked beech tree also is very common and on some sites it is by far the most abundant tree. Near the Delaware River in Camden County almost pure stands of beech trees have developed and here, as in North Jersey, the beech appears to propagate itself by root sprouts. Thus, it appears assured of a continued place in the forest canopy (Figure 12-1).

The tulip tree and the sweet gum also are abundant in some

of these upland forests, but because these two species do not develop well under shady conditions, both eventually may disappear or will certainly decrease in importance if the forests are left undisturbed. Red maple is common in both large and smaller tree classes. Several species of hickory, the black gum tree, and an occasional ash occur in the forests as well as representatives of a tree typical of more southern areas – the persimmon tree. In addition, the American holly tree, more typically a tree found on the sand dunes in New Jersey, is found also in the Mixed Oak forests on the more fertile soil in the western and southern parts of the Coastal Plains (Figure 12-1).

Unlike the Navesink Mixed Oak forest described earlier, there are typically no heath shrubs in the upland oak forests located in the western and southwestern part of the Coastal Plain. Instead there is an understory tree layer of dogwood, ironwood, and sassafras. Vines are common and include Japanese honeysuckle, Virginia creeper, and poison ivy. The spicebush, arrowwood, strawberry bush, and black haw are common shrubs in more moist locales. The most abundant herbs in this Mixed Oak forest are similar to those in the upland forest of North Jersey and include the may apple, jack-in-the-pulpit, false solomon's seal, and woodland asters.

Most of the forests on the Cape May peninsula are the lowland types described in previous chapters, though patches of the Mixed Oak upland type of forest also are present with representatives of beech and tulip trees. On some of the sandier soil here as well as that lying to the east of the Pine Barrens the oak forests also include a few trees typical of the sand dune forests described in Chapter 13 – the American holly and Spanish oak – and the sweetbay tree.

In his 1899 report on forests the state geologist noted that there were on the Inner Coastal Plain three groves of hemlocks which, as previously mentioned, are trees typical of areas to the north of New Jersey. The authors know of one such hemlock stand on a north-facing hillside just west of Hornerstown. It represents a good example of "relict" vegetation which in this case must have its origin at the time when glacier ice occupied

At left, Mixed Oak forest at Rancocas State Park showing a large black oak tree with American holly tree at its right. Note that the bark of the black oak tree is considerably darker than that of the white oak (*lower picture*). Blueberry shrubs are abundant in foreground. Lower picture shows Mixed Oak forest at Cheesequake State Park. ·White oak is larger tree at left and black oak is at right. Smaller trees are mostly red maple and chestnut oak. Laurel is abundant as shrub. *On opposite page,* the smooth gray-barked beech tree typically surrounded by root sprouts of seedling size. Picture taken on Inner Coastal Plain near Bordentown.

Figure 12-1 The Mixed Oak and Beech forests of the Mesic Coastal Plains Uplands.

northern New Jersey and the climate was cooler on the Coastal Plain.

Successional Plant Communities of the South Jersey Uplands

Many examples of the natural changes that occur in vegetation with the passage of time already have been cited. The uplands of South Jersey are no exception and if land is disturbed and then left idle without man's interference, various plant communities will occupy the land in succession. This type of situation most often has its origin in the fact that in New Jersey considerable acreage of land that once was farmed now is left idle.

On the uplands of South Jersey as in North Jersey the first plants to occupy an abandoned field are the herbaceous plants with short life-spans of one to two years. These are followed by longer living perennial herbs; a few seedlings of shrubs and trees develop into thickets and finally into a woodland. A study made of abandoned fields on the Inner Coastal Plain indicates that an annual called horseweed, is the plant that most frequently invades a field left idle; with it in smaller numbers grow other herbs including ragweed, corn chamomile, mouse-ear chickweed, prickly lettuce, red sand spurry, and foxtail grass. In fields left abandoned for ten to fifteen years, the most prominent plants are perennial herbs including goldenrod, broom sedge grass, and king devil hawkweed, but shrubs are also common – blackberry, dwarf sumac, red willow dogwood – and poison ivy is plentiful. The first trees to grow in the field abandoned on the Inner Coastal Plain are representatives of black cherry, sweet gum, red maple, and sassafras.

With the passage of only twenty-five years, an idle field on the Inner Coastal Plain may be transformed into a woodland consisting primarily of sweet gum and red maple trees as tall as 30 feet. Shrubs are not common in the woodland though some Japanese honeysuckle and arrowwood can be found and poison ivy is still abundant. Neither are herbs plentiful in the young woodland, though touch-me-not, goldenrod, bent grass, and in some places ground cedar may be common. As more years pass, the woodland develops into a mature upland forest of the type described earlier in this chapter. Inevitably, the change occurs and for this reason, if left undisturbed, the whole of the South Jersey uplands now so depleted of natural vegetation would be covered with forest.

In areas west and south along the borders of the Pine Barrens, almost pure stands of the Virginia pine tree occur as successional woodlands. This type of pine forest with its understory of heath plants, especially huckleberries, resembles forests in states to the south of New Jersey.

Vegetation of the Pine Barrens, the Drier Upland Habitat of South Jersey (Vegetation Type 11)

In the first decade of this century the Pine Barrens of New Jersey were described by Dr. Witmer Stone as follows:

The region is one of the largest in the Middle States in which anything like primeval conditions remain. Always sandy and thickly covered with more or less scrubby vegetation, interspersed with swamps and infested by hordes of mosquitoes, settlers have been in no hurry to clear it so long as more valuable land was available to the westward. Even to-day one may travel for ten or fifteen miles in some parts of the Barrens without seeing a habitation of any sort, and this within fifty and thirty miles respectively of New York and Philadelphia.

At the time that this was written the population of New Jersey was only one-third what it is today. Indeed, Dr. Stone observed that the advent of the automobile and other influences "are bound to make changes" in the region in the near future. It is hardly believable that in a book published more than fifty years later another writer, John McPhee, could accurately describe this region as follows:

From the fire tower on Bear Swamp Hill, in Washington Township, Burlington County, New Jersey, the view usually extends about twelve miles. To the north, forest land reaches to the horizon. The trees are mainly oaks and pines, and the pines predominate. Occasionally, there are long, dark, serrated stands of Atlantic white cedars, so tall and so closely set that they seem to be spread against the sky on the ridges of hills, when in fact they grow along streams that flow through the forest. To the east, the view is similar, and few people who are not native to the region can discern essential differences from the high cabin of the fire tower, even though one difference is that huge areas out in this direction are covered with dwarf forests, where a man can stand among the trees and see for miles over their uppermost branches. To the south, the view is twice broken slightly – by a lake and by a cranberry bog – but otherwise it, too, goes to the horizon in forest. To the west, pines, oaks, and cedars continue all the way, and the western horizon includes the summit of another hill – Apple Pie Hill – and the outline of another fire tower, from which the view three hundred and sixty degrees around is virtually the same as the view from Bear Swamp Hill, where, in a moment's sweeping glance, a person can see hundreds of square miles of wilderness. The picture of New Jersey that most people hold in their minds is so different from this one that, considered beside it, the Pine Barrens, as they are called, become as incongruous as they are beautiful.

Its distinctive flora, its wilderness, and its complete contrast with the urbanized surroundings make the Barrens a precious resource in the most densely populated state of the country. In size the Pine Barrens account for more than one-quarter of the state, 2,000 square miles, an area larger than all of Rhode Island and about equal to the Grand Canyon National Park. As mentioned earlier, the boundaries of the Pine Barrens are circumscribed by the particular soils underlying the region which are sandy and acid, of low fertility and with little moisture-holding capacity. Many plants cannot grow on this type of soil, but Dr. Stone reported finding 565 species of plants in the Pine Barrens, fifty-five of which were not known to grow elsewhere in New Jersey and two of which are not found anywhere else in the world. Since Dr. Stone made his study in 1910, additional plant species have been found and in a survey published in 1970 Dr. Jack McCormick raised to 800 the estimate of total Pine Barrens plant species. Dr. McCormick states that fourteen northern plants reach their extreme southern limit, or southernmost Coastal Plain limit, in the Pine Barrens. One of these is the curly grass fern mentioned in Chapter 8, a second is the broom crowberry discussed later in this chapter; others include the cotton grass and certain species of sedges, asters, and goldenrod.

Perhaps more impressive is the fact that 109 plants of southern affinity reach the northern limit of their range in the Barrens. The most commonly known of these include the turkey-beard, pixie moss, goat's-rue, American mistletoe, sand myrtle, bog asphodel, Pine Barrens gentian, golden-crest, and several species of asters, loosestrife, milkworts, goldenrod, violets, orchids, pipeworts, greenbrier, lobelia, grasses, rushes, and sedges. However, not all species are evenly distributed in the Barrens. Some grow only in the marshes, bogs, or swamps described in earlier chapters and others are more typical of the sandier and drier soil habitat that is the subject of this chapter.

Because the present vegetation of the Pine Barrens, like other parts of New Jersey, reflects the impact of past actions of man, it is appropriate to review the influences on the natural plant growth as background for understanding the present plant cover.

Past Influences on the Vegetation of the Pine Barrens

The many actions of man that have affected vegetation throughout New Jersey are described in Chapter 4. Of these, fire and cutting of timber have had the greatest impact on the vegetation of the Pine Barrens.

Fire as an environmental influence in the Pine Barrens probably was important even before man settled in New Jersey. For one reason, the coarse, sandy infertile soils of the region produce a vegetation that is highly flammable. Thus, fires caused by lightning will spread rapidly over this area particularly if there are no roads to act as firebreaks and if there are no fire-fighting activities. When Indians settled in the area, they set fire to the forests to facilitate hunting and travel; this practice was noted in the records left by European explorers and settlers. This combination of factors has meant that the vegetation of the Pine Barrens has essentially evolved in a fire environment. There are now no species in the upland vegetation that are killed by the fire; instead, when burned to the ground, the plants resprout readily from their underground systems. Some species, the lowbush blueberry among them, grow more vigorously following a fire.

It was in this context of a highly fire-susceptible background that the European settlers found in the Barrens extensive forests of tall pines with little undergrowth. Fires set by the Indians had been frequent enough to keep down dense shrub growth as well as to prevent the excessive accumulation of ignitable litter on the forest floor. The European colonists caused a great change in this forest, first by cutting down forest trees, and second by perpetuating the tradition of burning the woods.

The forests were cut initially to provide material for construction of houses and to furnish cordwood for heating homes in local settlements. But as the population expanded, enormous quantities of cordwood were taken from the region to provide fuel for those who lived in New York and Philadelphia and to supply power for locomotives and the early industries, particularly the bog iron industry.

Both for domestic use locally and for shipment to cities, charcoal was also produced. The imprint of charcoal production is left in the forest. To manufacture charcoal, wood was cut and heaped in circular domes as wide as 40 feet in diameter and as high as 20 feet. The piles were then covered with turf and sand removed from adjacent areas, then the wood was ignited and left to burn for about two weeks. At the end of this time the fire was smothered with a sand covering. The heat of the fire as well as the excavating of turf so destroyed the natural vegetation that areas with little or no vegetation still appear in the forest of the Barrens.

The demands on the Barrens woodlands lessened by the latter part of the nineteenth century. The termination of the bog iron industry and its successors—the manufacture of glass and paper products—combined with the substitution of coal for wood reduced the need to exploit the pine and oak timber of the region. With the industries gone from the area, the population diminished; villages were deserted and their ruins gradually became overrun by regrowth of Pine Barrens vegetation.

In sum, the influence of European settlement started by

cutting of the forest, which in the process greatly increased the burnable fuel on the ground as branches and tops of trees were left lying around. The fires that continued to burn through the area were more intense than those of Indian times because of the added fuel on the forest floor. Even now with the most modern equipment and methods and with a network of highways, it is difficult to control fires that start in the Barrens. As recently as 1963 almost 185,000 acres of forest burned on one weekend. The problem of fire is made complex by the fact that the longer the period during which fire is excluded from the Pine Barrens, the larger the amount of flammable vegetation collects and the more difficult it is to extinguish a forest fire once it starts.

Plant Communities of the Upland Pine Barrens Sites

The vegetation of the pine region is very diverse. One reason for this is the variation in soil moisture which provides different kinds of habitats for plants. The Barrens has a good representation of the wetter, lowland habitats of South Jersey — marshes, bogs, and swamps. The plant communities of these sites are described in previous chapters.

Diversity, particularly on the upland or drier parts of the region, is largely the result of man's activities through cutting and fire, although variations in soil characteristics from place to place also are influential. The natural tendency of the vegetation throughout most of the upland Pine Barrens region, were fire to be excluded, is toward the development of a strongly oak-dominated forest. This is because the young seedlings of pine trees cannot establish themselves when there is a substantial accumulation of litter (oak leaves, branches, etc.) on the forest floor, but such conditions are ideal for the growth of oak seedlings. Hence, in the absence of fire litter accumulates and young oak seedlings start to grow. As time goes by, old pines die and the oaks replace them as dominants of the forest.

On the other hand, when fire removes the litter on the forest floor, the exposed surface of the sandy soil creates an ideal condition for the establishment of pine seedlings. Once the pines grow beyond the seedling stage, they develop a reasonably thick bark

Mesic and Drier Habitats

Community Structure	Plants of Uplands Not Excessively Drained (Inner Coastal Plain Mostly)	
	Mixed Oak Forest	*Successional Stages of Vegetation*
Dominant Trees	White oak	*Stage 1-Herbs*
	Black oak	Horseweed
	Red oak	Ragweed
	Chestnut oak	Foxtail grass
	Scarlet oak	Goldenrod
	Beech (Dominant in areas)	Broom sedge
Other Typical Trees	Tulip tree	*Stage 2-Woody Invaders*
	Sweet gum	
	Red maple	Blackberry
	Hickory	Sumac
		Poison ivy
		Sweet gum
		Virginia pine
		Red maple
		Sassafras
		Tulip tree
		Wild cherry
Under-story Trees	Dogwood	*Stage 3-Woodland*
	Ironwood	Sweet gum (Dominant)
	Sassafras	or
		Virginia pine (Dominant)
		or
		Red maple (Dominant)
Shrubs	Spicebush	
	Others	
Herbs	May apple	
	Others	

Man's actions that affect vegetation

Figure 12-2   The Vegetation of the

in South Jersey

Plants of Sandy and Excessively Drained Flats (The Pine Barrens)			
Pine-dominated Forest	*Pine Plains (Dwarf) Forest*	*Oak-dominated Forest*	*Successional Stages of Vegetation*
Pitch pine	Pitch pine, stunted	Black oak (Dom.) Scarlet oak White oak Chestnut oak Post oak	*Stage 1-Herbs* Panic grass Horseweed Ragweed
			Stage 2-Woody Invaders
			Pines
Shortleaf pine Black oak Blackjack oak Post oak	Blackjack oak	Pitch pine Shortleaf pine Blackjack oak Sassafras	Oaks Sumac Sassafras Cherry Heaths
			Stage 3-Woodland Oak or pine
Bear oak Chinkapin oak	Bear oak	Few scrub oak	
Huckleberry Blueberry Other heaths	Crowberry Bearberry Many others	Huckleberry Blueberry Other heaths	
Wintergreen & others	Cow-wheat & others	Wintergreen & others	

Destroy vegetation or modify composition mostly by cutting and fire

Upland Habitats in South Jersey.

which is less susceptible to damage by fire than oaks of comparable age. Also, the two principal pines of the Barrens, pitch pine and the less abundant shortleaf pine, can send up new shoots from the base if the top of the tree is killed by fire. The result is that if fires occur, the tendency is to hold back the natural succession of vegetation in the Barrens to the pine forest stage.

Another important relation between the occurrence of fire and the composition of vegetation stems from the fact that different species within a single genus have different tolerances to fire. That the pitch pine and shortleaf pine are more tolerant to fire than oaks has been mentioned already, but even the oaks vary among themselves. On sites where fires occur at frequent intervals, the oaks occurring with the pines may be represented by only the post oak and the blackjack oak trees and where fires are very frequent, only the blackjack oak may be found. Among the smaller-growing (shrub) oaks the bear oak is the most tolerant of fire.

Thus, the occurrence of fire and especially its frequency determines in great part the extent to which one of three different plant communities prevails on the uplands of the Pine Barrens. We refer to these communities as the Pine-dominated forest, the Pine Plains (or Dwarf-pine community), and the Oak-dominated forest (Figure 12-2).

Pine-dominated Forest

The upland forest type usually associated with the Pine Barrens is one dominated by pine trees. It covers about 50 percent of the upland sites of the region. The most abundant tree and the one characteristic of the Barrens is the pitch pine, which in some places accounts for as much as 80 percent of the trees in the forest. It is not tall as trees go, usually reaching about 50 to 60 feet in height. The short leaf pine tree occurs in the forest but is present in smaller numbers. Black oak is one of the commoner oak trees in the Pine-dominated forest but one also finds white oak, post oak, scarlet oak, chestnut, and blackjack oak in varying proportions (Figure 12-3).

The shrubs which form almost a continuous understory under the trees are predominately members of the heath family. (See Appendix III, Part A). Of the latter, the black huckleberry and the lowbush blueberry occur almost universally and are joined occasionally by sheep laurel, fetterbush, and mountain laurel. Aside from the heaths there are shrub oaks, the occasional chinkapin oak and the more abundant bear oak. In some places the latter may be present in great numbers, making a very dense understory as much as 10 to 15 feet tall.

There are very few herbs present in the Pine-dominated forest and these occur very scattered throughout the forest. Little bluestem grass is one of the most common of these but one may find also representatives of wild indigo, turkeybeard, black oat grass, goat's-rue, cow-wheat, bracken fern, rockrose, and stemless lady's slipper. Actually, however, the herbaceous plants are so few when compared with the overwhelming abundance of woody plants that they make only a very incidental contribution to the vegetation.

The Pine-dominated forest in the Barrens is characteristically open, with trees growing sufficiently apart from each other to allow considerable light to penetrate to the lower layers. The crowns of the trees are also thin, intercepting less of the light than other conifer-dominated forests. For these reasons the image of the Pine-dominated forest presents a striking contrast to the Hemlock-Hardwoods forest (Figure 10-5) or the White Cedar bog forest (Figure 8-3), both of which are shady and dark in their interiors.

The Pine-dominated forest is favored by a fire frequency of about twenty years or somewhat less; and where fires have occurred more frequently, particularly as often as ten years or less, the Pine Plains type of vegetation has predominated.

The Pine Plains (or Dwarf-Pine Community)

A fire frequency of ten years or less appears sufficient to prevent the tree species that can stay alive under these circumstances from growing up to normal tree size. Instead, such trees

Above, as the Pine-dominated forest typically appears from roadside, Lebanon State Forest. *On opposite page,* a closer view of same forest showing bracken fern in the herb layer in foreground; scrub oak is growing to the left and behind the pitch pine trees in the foreground. Pitch pine is the most dominant tree in this particular forest but representatives of blackjack oak, post oak, black oak, scarlet oak, chestnut oak, and white oak also are present. Heath shrubs are common.

Figure 12-3   Pine-dominated forest.

224

grow as if they were low spreading shrubs and reach an average height of only about 4 feet. It is truly a "pygmy-pine" forest (Figure 12-4).

Only two tree species in the Barrens are able to recover from fires at intervals of ten years or less. These are pitch pine and blackjack oak, both of which after each fire start up again by sprouting from the root crown. Dr. Harold Lutz reported as many as 239 shoots growing from the root crown of a pitch pine in the plains. The pitch pine and the blackjack oak also produce seeds from the new basal shoots just a few years after a fire. Both the rapid sprouting after a fire and early seed production make these two tree species successful under the described conditions, but neither has a chance to grow up to normal tree stature before the recurrence of fire. As in the Pine-dominated forest, among shrubs it is the heaths—particularly huckleberry, and lowbush blue-berry—that are most ubiquitous. Mountain laurel is decidedly more abundant here than in the regular-sized Pine-dominated forest as are several other heath plants—trailing arbutus, bear-berry, wintergreen, sand myrtle, and sheep laurel. Other than the heaths the shrubby growth includes gallberry, false heather, sweetfern, and the tall-growing bear oak. The Pine Plains have much exposed ground between the tree sprouts and the shrubs; this is commonly occupied by mats of pixie moss which in early spring has showy white flowers.

It is in the Pine Plains also that the broom crowberry occurs. This, along with the curlygrass fern (Chapter 8), is considered one of two most unusual plants of the Pine Barrens. The broom crowberry or *Corema* has a northern range centered in the Mari-time Provinces of Canada; its presence in the Pine Barrens, the only location of the plant in New Jersey, represents the southern extreme of its range. Dr. Harshberger proposed that the vegeta-tion of the Pine Plains be called the Coremal community (pro-nounced to rhyme with "Chaparral"), the shrubby vegetation type that prevails in southern California and to which it bears such a remarkably strong resemblance. His suggested name was never adopted, however.

Until recently there was not general agreement as to the

Top: that the Pine Plains is truly a dwarf-pine community can be seen in this picture which shows the height of the vegetation relative to man. *Bottom:* blackjack oak (*lower left corner*) and pitch pine (*center*) in stunted form are the most abundant trees on the Plains; the common shrubs are the heaths.

Figure 12-4   Plains (or Dwarf-Pine) Community.

reason for the dwarf growth of the Pine Plains vegetation. Some felt that the stunted growth occurred only on particular types of soils; others suggested that strong winds might be responsible for the dwarf plants and still others proposed certain soil elements or even insects as the inhibiting influence on plant growth. Investigations have not supported any of these theories and now ecologists, foresters, and other scientists who have worked extensively in the Pine Barrens believe that the long history of frequent and severe fires is responsible for the unusual vegetation of the Pine Plains.

It is estimated that the Pine Plains now occupy about 15,000 acres. At present the acreage is not continuous but consists of three sections, the West Plains, the East Plains, and the Spring Hill Plains. Most of the West (or Upper) Plains, of approximately 6,000 acres, is located in Woodland Township of Burlington County, and State Highway 72 crosses its northern portion. The East (or Lower) Plains, also about 6,000 acres in size, is mostly contained in Bass River Township of Burlington County and State Highway 539 crosses the eastern part of it. The Spring Hill (or Little) Plains is a small area of only several hundred acres located 1 mile south of the West Plains in Washington Township.

Oak-dominated Forest

Oak-dominated forests compose about half the upland forest area of the Pine Barrens upland sites. Five species of oak trees commonly occur in this forest type – black oak, scarlet oak, white oak, chestnut oak, and post oak. The first four of these occur also in the uplands of North Jersey, but the post oak has a more southern range and in New Jersey it is primarily a tree of the pine region. Of these oaks, the black oak is generally the most abundant in the Oak-dominated forests; locally, however, chestnut oak, scarlet oak or white oak is dominant. The pitch pine and shortleaf pine both occur as scattered trees among the oaks.

The taller trees in the Oak-dominated forest are rather widely spaced but form a nearly closed canopy at a height of about 50 feet (Figure 12-5). Under the trees and forming a rather low

Figure 12-5   Oak-dominated forest of the Pine Barrens. While species of oak trees (particularly black oak) dominate the forest, occasional pitch pines can be found in it. The heath shrubs form a low understory layer. Picture taken in Lebanon State Forest.

shrub layer is a dominance of heath shrubs, primarily black huckleberry, dangleberry, and lowbush blueberry. Less common are the staggerbush, sheep laurel, and the smaller heaths — aromatic wintergreen and spotted wintergreen. There are very few herbs and those that do occur are the same species already mentioned as being sparsely present in the Pine-dominated forest.

Other Successional Plant Communities of the Pine Barrens

Although the Pine-dominated forests and the Pine Plains vegetation may be considered successional stages in the Pine Barrens resulting from cutting and fire, one also finds "old field" successional stages in the relatively few places where the land at one time was cleared for cultivation and subsequently has been

229

allowed to return to natural vegetation. These follow a successional pattern similar to that of the mesic uplands of South Jersey though with different plants. Horseweed and ragweed are abundant as early invaders. Several species of panic grass grow abundantly in the early stages of succession in the Pine Barrens. Shrubs and tree seedlings appear and soon an open woodland develops. Pitch pine is especially common as an early tree invader but along with it there may be oaks, wild black cherry, and sassafras. An oak forest or, depending on fire frequency, a fire-sustained pine forest, will ultimately develop.

The Pine Barrens and Man's Interference

In their concern to find practical means of getting maximum economic value from the forests of the Pine Barrens, foresters have adopted a technique called controlled burning. This technique involves essentially the practice established by the Indians who burned the forests at frequent intervals, and by certain cranberry growers who set fires in zones around their bogs to protect them from wild fires. In both cases, the results led to the development of Pine-dominated rather than Oak-dominated forests and also served to reduce the damage of wild fires by burning the excess dead material that litters the ground.

In controlled burning, foresters set fire to a forest in winter when fire can be controlled and when trees are cold so that the heat from burning litter is less apt to damage the tree; in this way the litter can be eliminated, dead stems and branches burned, shrub density reduced, and bare ground exposed (Figure 12-6). The practice creates the ideal open conditions for the establishment of pine seedlings and the maintenance of a pine forest, and is considered desirable because pine wood is more valuable than that of the oak in this area. Another benefit of burning in the cold season and under controlled conditions is that when the warm, dry, windy weather comes – the "fire season" – there is less fuel in the form of forest litter left to be burned. In this way devastating wild fires are minimized. It has been dramatically demonstrated that the practice of controlled burning can be used not

Figure 12-6 Controlled burning area in Lebanon State Forest. The burning is in an Oak-dominated forest area. The result is a pronounced reduction in the litter and the shrub layers; see comparison with forest shown in Figure 12-5 where no burning is done.

only as a tool to perpetuate the Pine-dominated forests but even to convert oak forests to pine forests.

Studies have shown that no species characteristic of the pine region is eliminated by controlled burning. This is as one would expect since the long history of fire in this area suggests a vegetation that has evolved in harmony with an environment in which fire is a part. In fact, if fire were eliminated in the Pine Barrens the diversity of plants would probably be reduced as the oaks increased at the expense of the pines. Such a circumstance would cause some species to become increasingly rare, and a few might even be eliminated from the Barrens entirely.

The growing public concern about the Pine Barrens is mentioned in a previous chapter dealing with the lowlands. Constant pressures threaten the destruction of large tracts of the drier uplands — the demand for Pine Barren land for use as an additional jet airport for the New York metropolitan area or for speculative

231

development schemes involving extensive industrial and residential use of the wilderness area. Conflicts arise between people who see the need to devote more land of the Pine Barrens to recreational activities of the growing population—canoeing, hiking, fishing, camping, picnicking, swimming, or hunting—and those who want to preserve as much land as possible in its natural state so as to protect the unique vegetation of the area. Forest management activities in the area—controlled burning and selective timbering—are not always understood or endorsed as desirable. More will be said in the final chapter about the pressing need to resolve these problems.

Summary

The natural vegetation of the mesic uplands of South Jersey is very different from that found on the drier soils of the Pine Barrens. The difference stems from the variation in soil types—the soil of the more moist uplands being more fertile than that of the Pine Barrens.

The present vegetation of the Pine Barrens is very diverse. Some part of the variety stems from the actions of man for it is believed that if the Pine Barrens were not exposed to frequent and intense fires, the Pine Plains type of forest (the Dwarf-pine vegetation) would not exist. In addition, if fires in the past had occurred less frequently, the Oak-dominated forest would prevail over the Pine-dominated forest.

REFERENCES AND SOURCE MATERIAL

Buell, Murray F. and John E. Cantlon. 1950. A Study of Two Communities of the New Jersey Pine Barrens and a Comparison of Methods. Ecology 31: 567–589.

Hanks, Jess. 1971. Secondary Succession and Soils on the Inner Coastal Plain of New Jersey. Bulletin of the Torrey Club 98: 315–321.

Harshberger, John W. 1916. The Vegetation of the New Jersey Pine-Barrens. Christopher Sower Co., Philadelphia, Pa. Reprinted in 1970 by Dover Publications, New York.

Little, Silas, Jr. 1946. The Effects of Forest Fires on the Stand

History of New Jersey Pine Region. Northeastern Forest Experiment Station, Forest Management Paper 2, Philadelphia, Pa.

McCormick, Jack. 1970. The Pine Barrens: A Preliminary Ecological Inventory. New Jersey State Museum, Research Report No. 2. Trenton, N.J.

McCormick, Jack and J. W. Andresen. 1963. The Role of Pinus Virginiana in the Vegetation of Southern New Jersey. New Jersey Audubon Society 110: 1–12.

McCormick, Jack and Murray F. Buell. 1957. Natural Vegetation of a Plowed Field in the New Jersey Pine Barrens. Botanical Gazette 118: 261–264.

McCormick, Jack and Murray F. Buell. 1968. The Plains. Pygmy Forest of the New Jersey Pine Barrens. A Review and Annotated Bibliography. New Jersey Academy of Science Bulletin 13: 20–34.

McPhee, John. 1968. The Pine Barrens. Farrar, Straus & Giroux, New York.

Moore, E. B. 1939. Forest Management in New Jersey. New Jersey Department of Conservation and Development, Trenton, N.J.

Phillips, J. J. and M. L. Markley. 1963. Site Index of New Jersey Sweetgum Stands Related to Soil and Water-Table Characteristics. U.S. Forest Service Research Paper NE-6: 1–25.

Stone, Witmer. 1910. The Plants of Southern New Jersey, with Especial Reference to the Flora of the Pine-Barrens and the Geographic Distribution of the Species. Annual Report, New Jersey State Museum, Trenton, N.J.

Thomas, Lester S. 1971. The Pine Barrens of New Jersey. New Jersey Department of Environmental Protection, Division of Parks and Forestry, Bureau of Parks, Trenton, N.J.

Vermeule, C. C., A. Hollick, J. B. Smith and G. Pinchot. 1900. Report on Forests in Annual Report of the State Geologist for 1899, Trenton, N.J.

13

Vegetation of the Coastal Sand Dunes

Introduction

The ocean coast of New Jersey lies entirely on the Outer Coastal Plain extending about 125 miles from the tip of Sandy Hook in the north to the southernmost end of Cape May. Sand dune vegetation occurs both on the mainland coast and on off-shore islands. To provide a framework for understanding the variety that occurs in dune plant communities, the chapter starts with an explanation of the processes by which our coastal features have developed. This is followed by a description of the vegetation of the sand dune formations and the chapter concludes with a review of the results of man's interference with dune vegetation.

Formation of the Coastal Features of New Jersey

Of all the land mass in New Jersey that of the coastal area is the most dynamic. Its contour changes noticeably from year to year and even from one day to the next. In a matter of hours during severe storms whole sections of beach in New Jersey have been wiped out. The sand carried away from one beach may be deposited at another or simply lost to the sea.

234

As mentioned in Chapter 2, opposing forces have been at work sculpturing the coastal area of New Jersey. Since the disappearance of the glacial ice about 12,000 to 15,000 years ago, the coastal land has been slightly uplifted, as a reaction to the removal of the tremendous weight of glacial ice. Concurrently, because of the melting of huge amounts of glacial ice, sea levels throughout the world have been rising. From about 14,000 until 7,000 years ago, the sea level rose relatively rapidly, over 3 inches in 100 years; since then there has been a lower rate of increase but worldwide levels are still rising about 1.5 inches each century. Thus, while the present coast of New Jersey shows features typical of coastal land that has been uplifted relatively recently, the ocean now covers areas that in glacial times had been dry land covered with vegetation. And even today the sea continues to encroach onto coastal land.

The New Jersey coast has many of the particular land formations that are characteristic of coastal land recently uplifted; these include barrier islands, spits, and hooks. It is primarily on these formations that the sand dunes are located.

Barrier Islands

Barrier islands are offshore sand ridges that parallel the shore and rise only slightly above high tide. These islands, also called offshore islands, are separated from the mainland by bodies of water referred to as lagoons. The seaside resorts of Atlantic City, Ocean County, and Wildwood are located on such islands (Figure 13-1).

The reason for the formation of barrier islands is in dispute. Some authorities believe the origin is explained simply by wave action on gently sloping coastal land that has been uplifted; others believe that tidal currents are more responsible for the formation than waves; and still others believe that both wave action and tidal currents are necessary for the development of a barrier island. Whatever its origin and wherever its location, a barrier island may progress naturally through a number of stages until it finally and completely disappears. Geologists sometimes

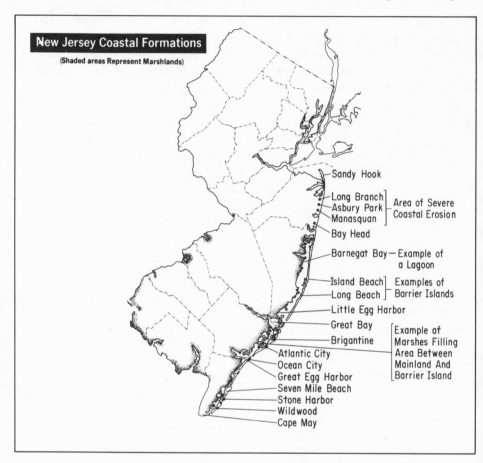

Figure 13-1   The coastline of New Jersey has land formations characteristic of coastal land recently uplifted – the offshore (or barrier) islands and the Sandy Hook formation. Along the southern coast, extensive tidal marshland is filling up the lagoons between the marshland and the offshore islands.

describe this process in terms of idealized stages of a life cycle – "early youth," "maturity," and finally "old age," the time at which the barrier islands completely disappear.

At the time of its formation a barrier island is separated from the mainland by an open body of water, but such a lagoon may rather quickly fill up and change its appearance. Tidal currents coming in through channel openings breached in the island carry sand which is then deposited in the lagoon. In addition, streams entering the lagoon from the mainland leave deposits of muddy sediments. As the lagoon becomes shallower, plants encroach outward on the water, as described in Chapter 7. Thus, the open body of water is transformed into marshland.

One of the forces influencing the change in the character of the offshore island formation in New Jersey and along other parts of the eastern seaboard is the rising sea level. This factor is less obvious to us than the impact of wave erosion, particularly at times of heavy storms. The wave action erodes away the sand on the barrier island, thus pushing the island back toward the mainland. Waves may finally win, and after years of erosion the barrier island may be pushed back so far that it becomes merged into the mainland and no longer is recognizable. Thus the mainland, which once was protected from storms by offshore islands, is left exposed to the ocean. Severe storms hasten the marine erosional process; storm waves have been known to exert as much as 6 tons of pressure per square foot on coastline rocks. The Jersey coast like the rest of the East Coast suffers particularly from exposure to waves created at the time of hurricanes and severe northeasters.

Various stages in the life of barrier islands are very evident on the New Jersey coast (Figure 13-1). On the part of the northern coast south of Sandy Hook and extending to Manasquan, the original barrier islands no longer exist. Islands that once existed in this area have disappeared. It is this area of the New Jersey coastal mainland that now is undergoing severe erosion because of its direct exposure to the ocean storms. The coastal strip between Long Branch and Manasquan particularly has suffered from the combined eroding action of waves and tidal currents and has lost a considerable amount of land in recent years. Some of the eroded sand has been carried northward by ocean currents and deposited at Sandy Hook, which as a result has been enlarging.

From Bay Head to Ocean City the barrier island formation is

still in its early youth stage of development and the lagoons of Barnegat Bay, Little Egg Harbor, and Great Bay remain open waters. However, below Bay Head there is a southward drift of sand that has caused changes in the location of the inlets connecting the lagoons or bays to the ocean.

According to Shephard and Wanless, authorities on world coastal formations, Barnegat Inlet since 1812 has shifted southward about a mile; considerable changes have also occurred in the Beach Haven and Little Egg inlet areas.

From Ocean City to Cape May, the barrier islands are in a later stage and marshland is now filling up the lagoons. The extent of the natural marshland is shown by the shading in Figure 13-1. Comparisons of photographs of the Cape May shore area show a considerable change in the coastline from 1944 to 1963. Man-made jetties that have caused accumulations of sand in some places—the Wildwood beach, for example—have hastened erosion of the coastal stretch between the city of Cape May and the southern tip of the cape. It is estimated that nearly a fourth of the land area of the old town of South Cape May has been eroded away in the last twenty years.

Altogether, it is estimated that the New Jersey barrier islands are retreating toward the mainland at an average rate of about 2 feet per year. This movement is accelerated by heavy erosion at times of severe storms. The effect of storms on the New Jersey coast has been reported by Dr. Kemble Widmer as follows:

In 1944, in four hours during a hurricane, and again in 1962 in two days of a northeast storm, a total of eight miles of New Jersey beach and dune at three locations, Harvey Cedars, Holgate, and Ludlam Beach, were completely washed away and swept clean and flat from the breaker zone - to the bay or salt marsh some miles inland. Extensive damage was caused elsewhere along the Jersey shore.

Spits and Hooks

The coastal feature called a spit is a ridge of sand attached to land at one end. Usually a spit parallels the mainland but it may curve inward toward the land and then is called a hook. Sandy Hook of the Jersey coast is an excellent example of this type of land formation (Figure 13-1). A spit or hook is built up by the deposition of sand carried by currents. Just since 1687 the northern end of Sandy Hook has been extended about 40 feet each year, more than quadrupling in size; this build-up of sand has been at the expense of beaches to the south. A lighthouse established in 1764 at the northern end of the hook was 1 mile away from the shore by 1930. However, the records since 1840 suggest that the rate of change in the profile of Sandy Hook is decreasing and in the more recent period the coastline has fluctuated only about 900 feet. Several times in the recent past waves have broken through the lower end of Sandy Hook, severing its connection to the mainland. An ocean sea wall was built in 1920 to prevent this, but inevitably waves will again sometime claim the land.

Sand Dune Formations

The sand dunes of New Jersey occur primarily on the barrier islands and on Sandy Hook, though there are a few dunes on the mainland, even as far inland as the Pine Barrens, where they are called hillocks. A dune is a windblown pile of sand, but when fully developed a dune formation has a particular topography which is well illustrated in New Jersey. The mature sand dune formations consist of more or less parallel and continuous ridges of sand with intervening troughs or depressions.

The dunes nearest the ocean are called primary dunes and the primary foredune is the particular dune ridge that lies immediately parallel to the water. Only the beach, an ever-shifting veneer of sand and pebbles, separates the primary foredune and the ocean. Normal high tides will cover all but the back part of the beach. Immediately behind the ridge of primary foredunes are flattened areas, hollows, and smaller dunes and this whole area is referred to as the primary backdunes. At Island Beach State Park the ocean beach is about 150 feet wide and the zone of the primary dunes (the area of the foredune and the backdunes) is about 450 feet in width. Additional dunes lie to the rear of the primary dune area and are referred to as secondary dunes. These, too, have a foredune ridge nearest the primary dunes and the more protected backdunes. Dr. William Martin, who did an extensive study of the vegetation of Island Beach State Park, describes the dune formations there as follows:

The primary foredune ridge is a straight and nearly continuous line of dunes immediately parallel to the shore. Broken by channels which vary in frequency from one place to another, it tends to be low (10–20 ft) in areas where the bar is wide and high (20–35 ft) in areas where the bar is narrow. The primary backdune area varies from a high sandy flat . . . to a conglomeration of dunes and hollows . . . to a broad, low swale between parallel foredune ridges. . . .

The secondary foredune ridge is a straight but discontinuous line of wide-spaced, lower, more stable, and apparently much older dunes centered on a line about 200 m [about 600 feet] from the ocean shoreline. The longest continuous segment of secondary foredune ridge . . . ranges from 20–25 ft elevation above mean sea level, is relatively broad, well-stabilized, and has gentle eastern (seaward) and western (landward) slopes.

From the standpoint of vegetation the distinctions between the primary and the secondary dune areas and within these, between the foredune and backdune areas, are important. Also, it is necessary to distinguish between the higher dune ridges and the hollows between the ridges. This is so because associated with these various topographic features are particular environmental conditions that influence vegetation—principally variations in the amounts of salt spray, sand movement, soil moisture, and soil salinity.

Wind-borne salt spray and sand movement are most intense on the ocean-facing primary foredunes which also have the least amount of soil moisture. Many plants are highly intolerant of salt spray, which may kill both plant foliage and new growth. The intensity of the spray decreases farther from the ocean as does the movement of the sand itself. Soil salinity is slightly higher in the more moist secondary inland dunes than on primary dunes. Dune ridges are very dry; hollows between the ridges have considerably more soil moisture. The relationship of these environmental gradients to vegetation types is discussed in the next section.

Natural Vegetation of the New Jersey Sand Dunes (Vegetation Type 12)

The coastal land of New Jersey includes two very diverse vegetation types – that of the sand dunes and that of the marshes. The latter is described in Chapter 7 and this chapter is devoted to a description of the plant cover found on that part of the sand dune area not covered by salt or fresh waters.

Complete botanical descriptions have been made of sand dune vegetation and at Island Beach State Park alone as many as 267 plant species have been identified, though almost half of these are species with only a few representatives. We identify herein four general types of vegetation that more or less typify the plants of the New Jersey sand dunes; these are the communities of the Dunegrass, the Beach heather, the Shrub thicket, and the Dune woodland. Each is described below, but it should be remembered that gradations occur among them, particularly among the last three (Figure 13-2).

Dunegrass Community

Few plants grow on the beach separating the primary dunes from the ocean, and those few occur in sparse numbers. In New Jersey the two most common beach plants are the sea rocket, an annual plant that flowers throughout most of the summer, and wormwood, a biennial plant which flowers only in its second (and last) year of life.

On the primary foredune, the sand ridge facing the ocean, only about 10 percent of the sand is covered with plants, but it is here that the first recognizable plant community occurs; this is the Dunegrass community, named for the most abundant plant. Dunegrass, also called marram or beachgrass, is a perennial plant with an extensive underground network that helps bind

Sand Dune Habitats in New Jersey

Community Structure	Dunegrass Community	Beach Heather Community	Shrub Thicket	Dune Woodland
	Mostly tall grass	Low cushion plant coverage	Shrubby thicket 1 to 15 feet high	Woodland
Typical Trees	None	None	Low growing Red cedar Black cherry American holly Scrub oak	Red cedar American holly Black cherry Pitch pine Hackberry Sassafras
Shrubs	None	None	Bayberry Beach plum Shadbush Blueberry	Bayberry Shadbush Blueberry
Herbs	Dunegrass (Dominant) Seaside goldenrod Sea rocket Spurge Beach pea Wormwood	Beach Heather (Dominant) and some Vines Virginia creeper Poison ivy Herbs Seaside goldenrod Others that grow in Dunegrass community	Vines of Virginia creeper Greenbrier Poison ivy	Vines of Virginia creeper Greenbrier Poison ivy

Man's actions that affect vegetation

Clear or Trample — Destroy vegetation and destroy dune formation

Figure 13-2　The Vegetation of Sand Dunes in New Jersey.

the sand and thus protect the dune from windblown damage. In addition, the dunegrass helps build up dunes by trapping additional windblown sand among its foliage and underground structures. Interestingly, it has been shown that there is a mutually constructive relationship between the accumulation of sand and growth of the dunegrass. Healthy dunegrass is most effective in slowing the movement of sand carried by the wind. Hence sand accumulates among the grass clumps and the dunegrass grows higher. At the same time the healthiest growth of the dunegrass is dependent upon a substantial rate of accumulation of sand.

According to Dr. Martin, on Island Beach the dunegrass accounts for as much as 82 percent of the sparse plant cover on the primary foredune. Growing with it in smaller numbers are other plants highly tolerant of salt spray. These include sea rocket, the large-flowered seaside goldenrod, dusty miller, and beach pea, and another type of wild grass known as beardgrass. Figure 13-3 showing this community type illustrates the dominance of dunegrass in the vegetation. The Dunegrass community grows throughout the primary dune zone as well as on the secondary foredune.

On certain Island Beach dunes the community is joined by a plant invader from Asia—a species of sedge (Carex kobomugi). Dr. John Small has postulated that the plant initially developed from seeds washed ashore from ships passing from the Orient.

Beach Heather Community

In the more protected areas of the primary and secondary dune zones, the low growing beach heather (also called beach heath) plant occurs in great abundance often forming a cushion mass over large areas (Figure 13-2). The beach heather being much less tolerant to salt spray does not compete with the plants of the Dunegrass community on exposed dunes. Rather, it tends to grow in the backdune areas at an average elevation of about 9 feet above sea level. This habitat is very dry but less exposed to salt spray and sand movements.

Natural branches of the beach heather spread across the

Figure 13-3   Dunegrass community on a primary foredune at Island Beach State Park. *Photograph by William E. Martin.*

dune, and the upright branches usually grow no more than 1 foot tall. In areas occupied by the Beach Heather community there is less bare sandy areas than in the Dunegrass community (Figure 13-3). The beach heather itself forms most of the plant cover in its community but growing with it though in sparse numbers are representatives of the herbs of the Dunegrass community mixed with horizontal vine networks of poison ivy and Virginia creeper and species of lichen, panic grass, and sedge. Examples of both the Dunegrass and Beach Heath communities can be seen on Sandy Hook and on most of the unsettled parts of the New Jersey barrier islands south of Manasquan.

Shrub Thicket Community

Inland from the ocean there is a decreasing amount of salt spray combined with an increase in soil moisture. These condi-

tions make for more varied natural plant cover including communities of woody plants. These occur mostly in the zone of the
secondary dunes, and even there, where the salt spray is the
highest, woody plants attain only low growth, forming more or
less a thicket. The most common plants in the Shrub Thicket
community include the bayberry, beach plum, shadbush, and
highbush blueberry shrubs with a few low-growing red cedar,
black cherry, scrub oak, sumac, and American holly trees. In
some places, extensive vines of poison ivy and Virginia creeper
combined with the thorny greenbrier make the thickets im-
passable (Figure 13-4). On the driest sites, the herb cover in the
Shrub Thicket community is sparse, with dunegrass and the seaside goldenrod most common. On moister sites several species of
ferns (particularly the royal, cinnamon, and marsh) and mosses
may be found. The height of the thicket vegetation varies from 1
to 15 feet depending upon the amount of salt spray reaching the
area—the taller heights being farther away from the salt spray
exposure. Isolated trees and shrubs in this area, if growing exposed to wind from the ocean, may take on a molded one-sided
form with branches on the lee side growing vigorously while
those on the windward side remain short, being killed back annually by wind-borne deposits of salt. Salt spray also inhibits
growth in height resulting in flattened tops as the plants reach
the level of the main windstream that crosses the barrier island.

Dune Woodland Community

In more moist and protected hollows of the secondary sand
dunes beyond the reach of heavy salt spray are the woodlands,
which are quite different from the mainland forests. At Sandy
Hook the Dune Woodland community is well known for its very
large American holly trees, some of which are more than 18
inches in diameter. Although the composition of the Dune Woodland community varies from site to site, the most common trees
on the Sandy Hook dunes include the holly, black cherry, red
cedar, red maple, pitch pine, hackberry, and sassafras. In places,
poison ivy, Virginia creeper, greenbrier, and a fourth vine, bitter-

Upper picture, Island Beach State Park with the beach heather in foreground and shrub thicket in background. *Photograph by William E. Martin. Lower picture,* Sandy Hook with beach heather and cactus plant in foreground.

Figure 13-4　The Beach Heather and Shrub Thicket plant communities.

sweet, grow abundantly and reach up to the treetops, making a dense forest (Figure 13-5).

The Dune Woodlands south of Sandy Hook are slightly different, for although more or less the same species occur, the proportions of abundance by species may change. For example, the woodlands on Island Beach are dominated mostly by red cedar, but in one small area the pitch pine tree is most abundant. In the red cedar woodland the associated trees include the holly, black cherry, sassafras, willow oak, and shadbush and the lower growing shrubs include bayberry, wild rose, blueberry, sweet pepperbush, and the usual vines—greenbrier, poison ivy, and Virginia creeper. Dr. Martin describes the red cedar woodlands of Island Beach as "junglelike masses of trees, shrubs, and lianas [vines]. The canopy is composed primarily of red cedar, and has an average height of 15–25 feet. . . . Shrubs and trees below the canopy are so intimately intermingled and interlocked with lianas that they present a virtually solid wall of vegetation."

The pine woodland at Island Beach is more open; the pitch pines growing between 15 and 30 feet high make up about half of the forest canopy trees. The white cedar tree is common and there are scattered representatives of American holly, blackjack oak, Spanish oak, white oak, and willow oak. Highbush blueberry and sheep laurel form much of the shrub cover in the woodland, but also present are some scrub oak, dangleberry, inkberry, bayberry, and greenbrier. Herbs are not common except in moist depressions where sphagnum moss and, in a few cases, sundew can be found.

The existing woodland at the Stone Harbor sanctuary is a lower growing forest composed chiefly of red cedar, holly, and black cherry trees. Unfortunately, there are few other dune woodlands remaining in New Jersey. In 1910 Dr. Stone reported both red cedar and pitch pine woodlands with composition similar to that described above on a tract of land south of Atlantic City. He also stated that until two years prior (1908) there had been "quite a wooded thicket" at the upper end of Ocean City comprising the same species except the pitch pine. At the time of his report Five-Mile and Seven-Mile Beaches were "thickly wooded"

Top, a red cedar woodland at Island Beach. *Photograph by William E. Martin. Bottom,* a red cedar woodland on Sandy Hook, and opposite page shows one of the large American holly trees growing on the sand dunes of the Hook.

Figure 13-5   Dune woodlands.

though he remarked that the forests were rapidly being leveled. From descriptions left to us, it appears that the former plant cover was similar to the present Island Beach red cedar woodland except that an additional tree, the swamp magnolia typical of more lowland sites (Chapter 9), also was abundant in the more southern dune woodlands.

In some depressions among the inland dunes, freshwater marshes may develop; the plant species occupying this type of habitat are similar to those described in Chapter 7. The secondary dunes extend to the bay side of the barrier islands where both saltwater and brackish marshes occur (also described in Chapter 7).

Man and the Vegetation of the Sand Dunes

As early as 1910, Dr. John Harshberger, a botanist who studied natural vegetation in New Jersey, wrote:

The development of the Atlantic sea coast from Maine to Virginia, and especially of the coastal lands of New Jersey, as places of summer resort has rendered a botanical survey of the shore line an imperative necessity. With the rise of towns and cities and the building of railroads, the primitive condition of the sea beaches has been remarkably changed. Dunes have been leveled, marshes have been filled in, old drainage areas have been removed, new soil has been brought to cover the sand formations to prevent their drifting, and these alterations have not failed to produce corresponding changes in the vegetation. New plants, weeds and the like, able, as well as the native plants, to withstand the saline conditions of air and soil, have been introduced with the coming of man as a permanent inhabitant; the old vegetation has been gradually removed, or, no longer able to grow under the altered conditions, has given place to the emigrants distributed by the aid of humans.

The fears of Dr. Harshberger were well founded and at present only in state park and sanctuary areas is the full variety of dune vegetation well preserved. One of the unusual dune woodlands described by Dr. Harshberger no longer exists. This was a forest at Wildwood on Five-Mile Beach in which he reported that "two monster hollies grow up to a height of sixty-five feet . . . a branch nearly a foot in diameter grew out of one tree and into the other solidly joining them together. . . ." In the same forest he observed huge red cedar trees nearly 3 feet in diameter, 50 feet high, and a grapevine 1 foot in diameter. In this forest many branches of the larger trees were said to be draped with a lichen giving the same appearance as the Spanish moss that grows on live oak trees in the South.

Ian McHarg, in a book entitled *Design with Nature,* describes eloquently the effects of man's abuse of sand dunes and its vegetation and the resulting retribution of nature. He calls to our attention what the people of the Netherlands have learned from their struggle to keep out the sea: sand dunes are stabilized by dunegrass, and stabilized dune ridges in turn serve as the primary defense against storm damage. These two facts combined with the knowledge that dunegrass cannot survive trampling by man lead the Dutch to allow only very restricted use of dune ridge areas. Following these same principles, McHarg suggests that for protection of its coast, New Jersey also restrict use of the offshore islands to preserve the natural barricade to the encroaching ocean. Beach areas would be open for recreation but for no building. In addition, no breaching of the ocean-facing or internal secondary dune ridge areas would be permitted. Rather, accessibility to beach areas would be by bridges over the dune ridges. The hollows between the ocean-facing and secondary dune ridges would be used primarily for recreation. Only the inland dune areas would be available for development of residential or commercial use and for highways. The bay shores would be left undisturbed.

There is considerable support for what McHarg says about the importance of dunes although a 25-foot high ridge of sand would seem like a pretty weak barrier to the ocean at times of the most severe storms; then, in fact, part of a dune ridge may be breached by the sea. Nevertheless, our coastal dunes are not just piles of sand; the buried network of living and dead underground parts of the dunegrass form a strengthening skeleton that can bind a dune together. Excessive trampling on dunes kills the living part of the system and thus reduces the strength of the dune and its resistance to storm damage. Hence restricted human activity on the dunes is important for their preservation.

Summary

The natural vegetation of sand dunes, the drier habitats of New Jersey, is determined primarily by environmental variations in soil moisture, salt spray, dune sand movement, and soil salinity. The ocean-facing primary dunes have the least soil moisture and the highest amount of salt spray and sand movement. Dunegrass lives successfully under these rigorous environmental conditions and while doing so serves to stabilize and to build up the accumulation of sand in the dune. Beach Heather communities occupy more protected areas of the primary sand dune. Taller growing shrub thickets and woodlands occur only in the more protected areas of the inland or secondary dunes. Man's disturbance of the natural dune vegetation results in the removal of the natural protective barriers for coastal land.

REFERENCES AND SOURCE MATERIAL

Chrysler, M. A. 1930. The Origin and Development of the Vegetation of Sandy Hook. Bulletin of the Torrey Club 57: 163–176.

Harshberger, John W. 1910. An Ecological Study of the New Jersey Strand Flora. Proceedings of the Academy of Natural Sciences, Philadelphia, Pa.

Martin, William E. 1960. An Unspoiled Bit of Atlantic Coast. Journal of the American Museum of Natural History 69: 8–19.

Martin, William E. 1959. The Vegetation of Island Beach State Park, New Jersey. Ecological Monographs 29: 1–46.

McHarg, Ian L. 1969. Design with Nature. The Natural History Press, Garden City, N.Y.

Milliman, John D. and K. O. Emery. 1968. Sea Levels During the Past 35,000 Years. Science 162: 1121–1123.

Shepard, Francis P. and Harold R. Wanless. 1971. Our Changing Coastlines. McGraw-Hill, New York, New York.

Small, J. A. 1961. The Vegetation of the Seacoast of New Jersey. New Jersey Nature News, New Jersey Audubon Society 16: 51–58.

Small, J. A. and W. E. Martin. 1958. A Partially Annotated Catalogue of Vascular Plants Reported from Island Beach Park, New Jersey. Bulletin of the Torrey Club 85: 368–377.

Stone, Witmer. 1910. The Plants of Southern New Jersey, with Especial Reference to the Flora of the Pine Barrens and the Geographic Distribution of the Species. Annual Report, New Jersey State Museum, Trenton, N.J.

Widmer, Kemble. 1964. The Geology and Geography of New Jersey. Van Nostrand Co., Princeton, N.J.

14

New Jersey Forests as Part of the Eastern Deciduous Forest Formation

Introduction

The natural vegetation of the world has been classified in broad categories called formations. When mapped, the boundaries of formations appear to coincide more or less with those of distinct climatic regions. In North America one of the major formations of natural vegetation is called the Eastern Deciduous forest, the classification to which the forests of New Jersey belong.

This chapter starts with a brief summary of the major formations of vegetation in continental United States and Canada. The position of the New Jersey vegetation within the Eastern Deciduous forest formation is then discussed.

Major Formations of Vegetation in United States and Canada

Although ecologists and plant geographers are able to differentiate among major categories of vegetation occurring throughout the world, they find it difficult to map the boundaries of each group accurately. This is so for several reasons. First, in the real world there are seldom sharp separations that distinguish the end of one plant formation and the beginning of another. Rather, more often than not there is a transitional zone between the two formations combining some elements from both plant groups. Subjective interpretation then is the only basis for delineating specific boundaries of the plant formations. Adding to this complexity is the fact that climatic elements, the primary cause for differences in vegetation formations, vary not only with differences in longitude, latitude, and geographic position relative to seas but also with altitudinal differences within a specific geographic region. Nevertheless, classification of vegetation into formations and smaller groups within these major categories, however arbitrary, serves important purposes. For one, it provides a basis from which causes of vegetation differentiation can be analyzed.

Simplified in the sense of number of formations and in delineation of their boundaries, the major formations of vegetation in continental United States and Canada are mapped in Figure 14-1. Eight types of vegetation are so distinguished; these are as follows:

1. *The Tundra formation* is a very low growing type of plant growth; many of the plants are lichens — particularly abundant is one called reindeer "moss." With the lichens grow grasses, sedges and perennial herbs, some of which produce large colorful flowers in July and August. Only a few woody plants are found on the tundra and these are dwarfed or grow in prostrate form as in the case of birch or willow trees. In North America the tundra forms a broad band across the northern part of the continent, separating the permanent ice and snow masses in the far north from the northern forests to the south. Tundra type of vegetation also is found above timber line on mountains of higher elevation as far south as Mexico.

2. *The Northern Conifer forest formation* is a forest type of vegetation in which most of the trees are evergreen conifers — trees that bear seeds in cones and that have needlelike leaves that remain on the tree all year round. Especially abundant are

Major Formations of Vegetation in United States and Canada

Key to Formations shown on Map

1. Tundra
2. Northern Conifer Forest
3. Eastern Deciduous Forest
4. Grasslands
5. Western Conifer Forest
6. Desert
7. Chaparral
8. Sub-Tropical Forest

Figure 14-1   Eight major formations of natural vegetation in United States and Canada. Map adapted with changes from Henry Gleason and Arthur Cronquist. The Natural Geography of Plants. Columbia University Press. New York, 1964.

various species of spruce and fir trees which are joined in some areas by large numbers of the white (or paper) birch, trembling aspen, balsam poplar, and various species of pine trees. The Northern Conifer forest formation, also called the Boreal forest, lies south of the tundra in Alaska and Canada, and extends down into the higher elevations of northern New York State and New England.

3. *The Eastern Deciduous forest formation,* also a forest type of vegetation, is one that grows south of the Northern Conifer forest in the Eastern United States. The formation derives its name from the fact that in the climax forests of the region the conifer trees are outnumbered by the so-called deciduous trees which have broader leaves that fall off the tree at the start of the cold season. More is said below about the composition of this formation.

4. *The Grasslands formation* is a low-growing type of vegetation typical of the drier midwestern areas of United States. The plants most abundant in this formation are various species of grasses, the taller of which grow in the areas of the most rainfall, or the prairies. Distinctively different in appearance are the plains—the drier grassland areas—which are covered by lower growing grasses. Sites with moisture conditions between the prairies and those of the plains are covered by representative species from both groups as well as grasses of intermediate stature. The three divisions of the grasslands form roughly north-south oriented zones; the tall-grass prairie zone is in the east, the short-grass plains are in the far west, and the mixed-grass zone lies between the two.

5. *The Western Conifer forest formation* includes a number of different forest types but all are dominated by conifer trees. These forests grow on the mountain areas of western United States and Canada including the Coastal Ranges, Cascades, Sierra Madres, Sierra Nevadas, and the Rockies. Variations in the composition of the forest occur with altitudinal changes but throughout various species of spruce, fir, and pine are the most abundant trees. Forests on the West Coast also include in predominant numbers the giant redwoods and western species of hemlock and red cedar.

6. *The Desert formation* includes the type of vegetation that grows under the driest of conditions. Located between western mountain ranges, the deserts receive little rainfall; only specialized plants can survive where soil water is in such low supply.

Many desert plants exhibit structural adaptations that permit their survival under dry conditions, such as the ability of the cactus plant to store water in its succulent tissues. Sagebrush, the creosote bush, bur sage, and shadscale are among the more abundant shrubs of the desert. A surprising number of showy-flowered annual herbs develop and flower in the desert in the short intervals of rainfall in spring and late summer.

7. *The Chaparral formation* is a type of vegetation peculiar to lower California. The distinguishing plants in this formation are evergreen shrubs growing 3 to 10 feet tall and with rather small leaves that remain on the stems all year. Typical of these are the chamiso shrub, numerous species of manzanita, and another shrub known as California lilac. The shrubs dominate the vegetation, forming dense thickets, though in cooler and moist areas they are joined or even replaced by evergreen trees such as the tanbark oak, the California laurel, the giant chinquapin, and the madrone.

8. *The Subtropical forest formation* occurs in the continental United States only at the southernmost tip of the Florida peninsula and in the Florida Keys where, in swamps and on higher hammocks, there is natural growth of some species of palms such as the sabal palm as well as the red mangrove, the wild tamarind, the gumbo limbo, the strangler fig, and other broad-leaved evergreen trees representative of more tropical climates.

**Eastern Deciduous Forest Formation
and Its Subdivisions**

As shown on the map, the Eastern Deciduous forest formation covers all the eastern United States except southern Florida; in area it includes more of continental United States than any other formation. Its boundaries are delineated for the most part in the north by colder temperatures, on the west by lower rainfall, and on the south by higher temperatures. For the most part, the formation comprises a number of different types of deciduous forests, though conifer forests are found on the higher elevations in the northeast and on the coastal plain area in the southeast.

The deciduous trees in the Eastern forest formation grow

normally no more than about 100 feet high and rarely reach the spectacular heights of their counterparts in the forests of the western United States. The beautiful fall displays of leaf colors that are seen in the eastern forests, however, are absent in the West.

The area circumscribed for the Eastern Deciduous forest formation includes the most densely populated and highly developed part of the country (Figure 14-1). Virgin forest once covered all the upland sites but most of it has been destroyed. In New Jersey, and on the East Coast in general man not only has displaced much of the natural vegetation but also through his past actions has modified the composition of much that remains. Nevertheless, it is still possible to recognize distinctive subdivisions of the Eastern Deciduous forest formation, each a region of potentially distinctive climax forest types. The adjective "climax" as used here is defined in Chapter 1 and means forest types which are relatively stable or permanent in time as contrasted with "successional" forest types which have a relatively short duration on a particular site.

Many attempts have been made to delineate acceptable subdivisions of the Eastern Deciduous forest formation; the classification most widely accepted though with some qualifications is one developed by E. Lucy Braun. In her book, *Deciduous Forests of Eastern North America,* Dr. Braun delineates nine regions of potentially different climax forest types and then further subdivides each region into varying numbers of sections. The names of the nine regions whose boundaries are mapped in Figure 14-2 are derived for the most part from the identity of the most abundant climax trees in the region. They are as follows:

1. *The Hemlock-White Pine-Northern Hardwoods forest region* is the forest climax type of the northern part of the Eastern Deciduous forest formation extending from New England westward to northern Wisconsin and Minnesota.

2. *The Oak-Chestnut forest region* includes southern New England and New York, northern New Jersey, and parts of Pennsylvania, Maryland, Virginia, North Carolina, and eastern Tennessee.

3. *The Oak-Pine forest region* lies to the east and south of the

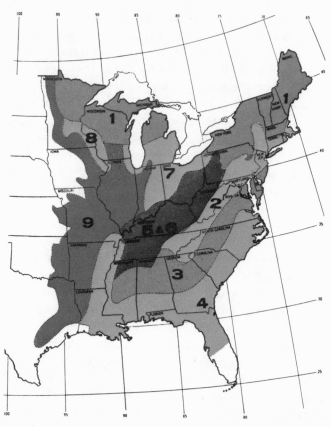

Figure 14-2   The Braun classification of forest types in the Eastern Deciduous Forest Formation. Map redrawn from E. Lucy Braun. Deciduous Forests of Eastern North America. Hafner Publishing Co., New York, 1950.

Oak-Chestnut forest area and extends from southern New Jersey to inland parts of southern states as far west as Mississippi.

4. *The Southeastern Evergreen forest region* comprises the area of the southern coastal plain. The forest name is derived from the fact that the most abundant trees of the region are various species of pine trees with evergreen needles. However, many ecologists believe that if the area were protected from fire, the evergreens would be replaced by a mixture of hardwood trees — that is, deciduous trees such as oaks and hickories.

5. and 6. *The Mixed Mesophytic forest region and the Western Mesophytic forest region* have numerous tree species growing in varying mixtures, and no one or several trees can be used to identify the forest types. The boundaries of the two regions collectively coincide with the Allegheny and Cumberland mountains and plateaus extending to the bluffs of the Mississippi River in the west.

7. *The Beech-Maple forest region* lies mostly in three states — Ohio, Indiana, and southern Michigan.

8. *The Maple-Basswood forest region* in area is the smallest forest region of the Eastern Deciduous formation and it covers parts of Minnesota and southern Wisconsin.

9. *The Oak-Hickory forest region* forms the whole western border of the formation extending from Minnesota in the north to Texas in the south.

Braun's classification scheme just described has some obvious shortcomings with respect to its treatment of New Jersey. For one, because of the die-off of the chestnut trees, the category entitled "Oak-Chestnut forest region" has long since been a misnomer when used for the forests of northern New Jersey as well as those in the remaining part of the region.

Another classification scheme has been suggested recently for the forests of the Northeast. This was developed by the Northeastern Division of the U.S. Forest Service with cooperation of state foresters and mapped in a publication by Howard Lull. Under this classification, the commercial forest land in the Northeast (from Maine to West Virginia and Maryland) is classified by five regions. Two of these are confined to northern New England and New York — the Spruce-Fir Forest and the Beech-

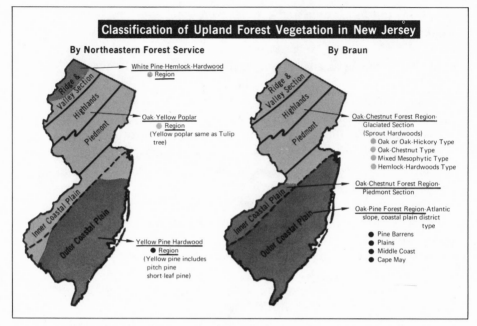

14-3   Two classifications of upland forest vegetation in New Jersey; one made by the Northeastern Forest Service and the second by Dr. E. Lucy Braun. Maps were prepared from data in Howard Lull, A Forest Atlas of the Northeast, 1968, and E. Lucy Braun, Deciduous Forests of Eastern North America, 1950.

Birch-Maple Forest. The remaining three forest regions extend more southward and some part of New Jersey is included in each (Figure 14-3). The three regions are as follows:

1. *The White Pine–Hemlock–Hardwood region* extends from the southern part of Maine through Massachusetts and northern Connecticut to southern New York State and northern Pennsylvania. Only the tip of northwestern New Jersey (the northern part of Sussex County) is included in the region.

2. *The Oak–Yellow Poplar (or Tulip Tree) region* includes the remainder of northern New Jersey and the area of the Inner Coastal Plain, along with southern Pennsylvania, West Virginia, and northern Delaware and Maryland.

3. *The Yellow Pine–Hardwood region* comprises the Outer

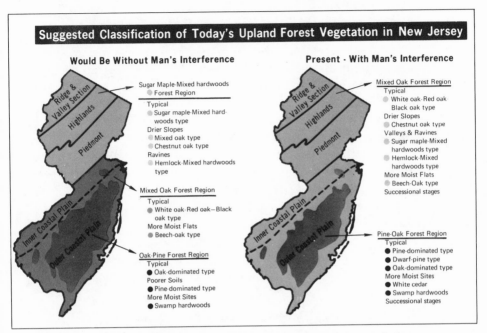

Figure 14-4   The authors suggest the dual classification mapped above that distinguishes between the present forest regions and those that would exist without man's interference. Such a classification highlights the influence of man's actions on natural vegetation.

Coastal Plain of New Jersey and southern Delaware and Maryland. The name yellow pine as used here includes the pitch pine and shortleaf pine as well as other species of pines that grow in states south of New Jersey.

The names given to these major forest regions appear somewhat inappropriate as descriptions of climax forests in New Jersey alone, though in the supplementary descriptions given of each forest region the particular types of forests typically found in New Jersey are included. However, because in New Jersey the tulip tree is only a successional tree unable to develop well under the shady conditions in a mature forest, the use of its name to describe a climax upland forest is inappropriate. The reason that foresters emphasize the presence of this tree is its great value as timberwood.

Suggested Classification for New Jersey Forests

The forest classification schemes reviewed in the previous section point up two needs to be satisfied for successful description of the forest vegetation of any region. First, a distinction must be made between classifications which depict the prevailing forest types as they are today and those which suggest what the climax forest types might be without the interference of man. Second, within any classification scheme of upland forests, some allowance must be made for variations in composition because of somewhat drier or wetter soil conditions.

Following these principles a scheme for a dual classification of New Jersey upland forest vegetation is presented in Figure 14-4. In one classification, the present prevailing upland forest regions are delineated, and in the second, the types of forests as they might exist without man's interference are categorized.

The New Jersey Forests as They Now Exist

It is suggested that the upland forests of New Jersey as they presently exist should be classified within the Eastern Deciduous forest formation as belonging to two regional categories:

1. *The Mixed Oak forest region,* as shown on the map in Figure 14.4, includes all of New Jersey except the area of the Pine Barrens. The typical upland forest in this region has as its most abundant trees the white oak, red oak, and black oak in varying proportions. This is the forest type that now prevails on the mesic uplands of North Jersey (Chapter 10) and on the mesic uplands of the Coastal Plains of South Jersey (Chapter 12). On the drier slopes of North Jersey, the Chestnut Oak type predominates (Chapter 11) and in more moist valleys or ravines of North Jersey either the Sugar Maple–Mixed Hardwoods forest or the Hemlock–Mixed Hardwoods forest (Chapter 10) is found. On the more moist flats of areas in the Inner Coastal Plain, the Mixed Oak forest contains abundant numbers of the beech tree (Chapter 12). Successional stages of vegetation also are present throughout the region (Chapters 10, 11, and 12).

2. *The Pine-Oak forest region* is the best broad regional heading for the present forest vegetation of the Pine Barrens. It would encompass the Pine-dominated forest type, the Pine Plains type, and the Oak-dominated forest type, all three of which were described in Chapter 12. On the moister sites in the region, the White Cedar or the Swamp Hardwoods lowland forest types prevail (Chapter 9). Successional stages of vegetation also are common in this region (Chapters 9 and 12).

The New Jersey Forests as They Would Be without Man's Interference

If man had never settled in New Jersey, the upland vegetation today would be quite different. In this case it is likely that the forests blanketing the state would be classified as belonging to three rather than two regional subdivisions as follows:

1. *The Sugar Maple–Mixed Hardwoods forest region* would include all the mesic uplands of North Jersey, and the prevailing forest type would be that of the Sugar Maple–Mixed Hardwoods described in Chapter 10. The Mixed Oak forest type and the Chestnut Oak type would be found only on the drier slopes of the region; the Hemlock–Mixed Hardwoods forest (Chapter 10) would occupy the same types of sites (ravines and north facing slopes) that it occupies today.

2. *The Mixed Oak forest region* would be substantially reduced in size and would include only the area classified as the South Jersey mesic uplands which is the Coastal Plain area exclusive of the Pine Barrens (Chapter 12). Without the settlement of man, the Mixed Oak forest types which now occur only as small pockets of natural vegetation would cover the landscape.

3. *The Oak-Pine forest region* would be a better name for the area now known as the Pine Barrens if it were not for the history of fire and cutting because on all but the poorer soils the Oak-dominated forest type would prevail (Chapter 12).

The dual forest classification that distinguishes between present forest regions and those that would exist without interference highlights the influence of man's actions on natural vegetation. Man has completely destroyed natural plant growth; what is less widely understood, however, is his ability to alter the composition of the natural vegetation that remains. He has the capability of making the natural landscape more or less diverse as well as the capability of obliterating it completely. For example, the forests in the northern part of the state actually could become more diverse if for a long period of time fire and woodland cutting were eliminated. On the other hand, to maintain the present diversity of forest types in the Oak-Pine region of South Jersey, controlled burning and cutting of woodlands by foresters is necessary.

Summary

The forests of New Jersey are but a small part of the Eastern Deciduous forest formation, one of the major categories of natural vegetation in North America. The classification of the New Jersey forests within its formation group is difficult because man's actions have transformed the composition of the forests. For this reason, a dual classification scheme is suggested, one to describe the forest vegetation as it is today and another to provide a classification scheme of the forest types as they would be without the interference of man. A comparison of the two classifications highlights the capability that man has to increase or to decrease landscape diversity by altering the composition of natural vegetation.

REFERENCES AND SOURCE MATERIAL

Braun, E. Lucy. 1950. Deciduous Forests of Eastern North America. Hafner Publishing Co., New York.

Gleason, Henry and Arthur Cronquist. 1964. The Natural Geography of Plants. Columbia University Press, New York.

Lull, Howard W. 1968. A Forest Atlas of the Northeast. Northeastern Forest Experiment Station, Upper Darby, Pa.

Quarterman, Elsie and Catherine Keever. Southern Mixed Hardwood Forest: Climax in the Southeastern Coastal Plain. U.S.A. 1952. Ecological Monographs 32: 167–185.

Part V
A Look into the Future

15

The Future: Vegetation and Man in New Jersey

Introduction

As stated in Chapter 1, one purpose of this work is to describe the present natural vegetation of New Jersey and another is to explain why the vegetation is what it is.

It is hoped that from previous chapters the reader has derived a sense of the unusual diversity that exists in the vegetation of New Jersey. That the most densely populated state in the union has any natural vegetation at all probably comes as a surprise to some people, especially those acquainted only with that part of the New Jersey landscape seen en route from New York to Philadelphia. And that the vegetation exhibits such a wide variety may have come as an eye-opener even to those more familiar with the state.

Understanding of the vegetation and of the reasons for its diversity starts with a recognition of the distinctive plant habitats found in the state. Twelve of these were identified and the characteristic vegetation of each was described in preceding chapters. One's understanding of the present natural plant growth is not complete, however, until the impact of man's activities on the landscape is fully realized.

If this work helps the reader to recognize and to understand better the natural vegetation that is now present in New Jersey, it has served its first purpose. But the total objective is more than this. Although a fuller appreciation of the natural landscape in itself is of value, a greater accomplishment would be to arouse more widespread interest in and concern for the future environment of New Jersey. To express it differently, one aim of this book is to convey to its readers a basic understanding of why the vegetation of New Jersey is what it is today and so to stimulate them to consider seriously, and in ecological terms, the nature of the landscape that should remain for future generations. The purpose of this chapter is to encourage such interest by highlighting the conflicts and issues that inevitably will arise in New Jersey with respect to existing areas of natural vegetation.

A Source of Conflict – The Fear of "Running Out of Space"

A decade ago when Jean Gottmann wrote *Megalopolis,* he noted that in many quarters of the northeastern United States there is a deep-seated fear of "running out of space." Admitting that the term is a vague one, he offered some interpretations of its meaning. For one, to run out of space could mean that a whole area might be so crowded that "people would not be able to move freely about it, and there would be no choice left as to where to live, how to live, what to do for recreation, where to work." In this sense, then, "running out of space" means lacking free access to desired places. Gottmann suggests another meaning for the term – a lack of nearby access to "open space, especially to what the average citizen wants for recreation."

Since the publication of Gottmann's book in 1961, the population of New Jersey has increased by one-sixth. Now, with an average of nearly 1,000 people per square mile, the population density of New Jersey surpasses that of India, Japan, and the Netherlands, areas normally considered overcrowded. For this reason forecasts of additional population growth in the state can be somewhat frightening, especially since past projections have been too conservative and actual growth has topped the expectations. It is estimated that by 1980 the population of New Jersey

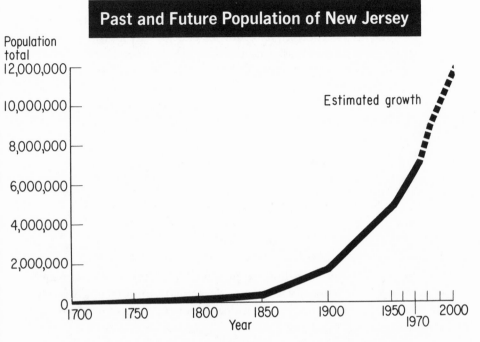

Figure 15-1   Official estimates of population growth project a figure of twelve million residents in New Jersey by the year 2000. This would mean two-thirds more people in the state than in the year 1970. Source of data: 1969 Population Estimates for New Jersey. New Jersey Department of Conservation and Economic Development, Bureau of Research and Statistics.

will reach almost 9 million, a total that is 25 percent greater than the 1970 population. By 1990 the number of people will be half again as many as in 1970 and, if official estimates are accurate, by the year 2000 the population of New Jersey will be about 12 million, a figure two-thirds greater than that of 1970 (Figure 15-1).

The result of this predicted population expansion and accompanying economic growth, if it does actually occur, will be simple and direct: vast amounts of open areas now covered by natural vegetation will have to be appropriated for housing and for commercial and industrial buildings to accommodate and to serve the increased numbers of people and to provide them with employment. Thousands of miles of additional highways will be needed to alleviate already pressing traffic congestions. The ultimate effects of converting present open areas to man-made artifacts are not easy to predict; and to gain some perspective into the problems, it is necessary first to pinpoint some of the benefits now served by areas of natural vegetation.

Physical Benefits of Natural Vegetation Areas in New Jersey

The efforts of zealous conservationists to preserve wilderness areas in the Far West have led to some confusion about the need to retain areas of natural vegetation in the more heavily populated regions of the country. For example, in the sense of retaining vegetation in its natural pristine state, "preservation" has no meaning in New Jersey, where practically no virgin growth remains. Justification for retention of natural areas therefore must rest on other grounds.

Disregarding for the moment social values, which will be discussed later, there are tangible physical benefits that accrue from the areas of natural vegetation. Many of these have been previously mentioned in connection with the description of the various vegetation types of the state. First, plants, in the process of photosynthesis, perform two essential, priceless, and exclusive functions without which man could not exist—these are the conversion of nonliving material to food and, as a by-product of this activity, the release of oxygen to the earth's atmosphere. But in addition each natural plant habitat in North and South Jersey, whether it be a marsh, bog, swamp, floodplain, forested upland or slope, or even a plant-covered sand dune, serves man in other ways.

The real worth of marshland only recently has gained widespread attention. The marshes serve as the spawning and nursery grounds for many kinds of shellfish, including clams, crabs, shrimp, oysters, and scallops. Bluefish and weakfish are only two of the many fin fish whose continued propagation depends upon the availability of protected estuary zones. Thus, if New Jersey is to continue to have an opportunity for commercial and recreational deep sea fishing, preservation of the remaining marshland is essential.

Marsh vegetation as well as that of other lowlands also serves as a migratory shelter or a nesting habitat for many different kinds of waterfowl and other birds. The names for Great Egg and Little Egg harbors were derived from the fact that each year thousands of ducks, geese, and swans used to nest along the shores. This has been sharply reduced by man's destruction of so much of the natural habitat in the area. The numbers and also the variety of other wildlife have diminished as marshes and other lowlands have been obliterated by fill-in or by drainage or polluted by toxic wastes. Mink, otters, muskrats, and raccoons are only a few of the animals that once were far more common than now in the natural wetlands of New Jersey.

Areas of natural wetlands are important as protectors against flooding and erosion of higher land. Coastal marshes protect the landward bay areas against high tides and damage from storm waves. The stems and roots of the salt marsh vegetation blunt the force of the tide waves and curtail erosion of the land. It is of equal importance that tracts of undeveloped coastal marshes act as water storage facilities to absorb incoming floodwaters. Water levels can rise in these lowlands without causing any destruction, for a rise of only inches in an area as small as 10 acres of marshland will store more than a million gallons of water.

In the same way, if left undeveloped, floodplain forests on the lowlands bordering inland streams serve as protectors against flood damage from swollen rivers (Figure 15-2). Floods tend to occur in inland New Jersey in spring when fast melt of snow accumulations in the northern part of the state adds abnormally large volumes of water to upland river sources. At other times of the year very heavy and concentrated rainfall may add more water to rivers than the stream banks can contain. The simplest and most effective way to moderate the intensity of flood

Figure 15-2   The flood plain of the Millstone River on April 6, 1957, when torrential rains caused the river to overflow its banks. This has happened several times just within the last five years, and if this flood plain had been developed, the flooding would have caused severe property damage.

damage in either case is the preservation of natural floodplain forests along river courses. Such areas act to dampen the temporary excess surges of water. In places where these temporary water storage areas have been eliminated because floodplain forests have been cleared or marshlands have been filled in, the floodwaters are less restrained and hence more likely to overrun settled areas causing extensive property damage and, in some cases, loss of life. On August 28, 1971, a tropical storm called Doria caused the worst floods in the history of Central New Jersey. In Mercer County alone the damage was estimated at 30 million dollars. Not only were residential developments, industrial plants and transportation facilities on lowlands heavily damaged but in adjacent areas basements were flooded, roads were washed out, and water supply, sewer and power systems were made inoperative. The extent of the damage caused by the storm resulted in a much greater appreciation and understanding of the need to restrict development of the state's lowlands.

In addition to acting as a mechanism for flood control, natural lowland areas also serve as a source for the replenishment of groundwater supplies. Although some of the water that floods lowlands is lost to the air either by evaporation or by plant transpiration, a sizable amount often sinks into the ground and becomes part of the underground water supply. Where water is in short supply, as in New Jersey, the maintenance of large reservoirs of groundwater is extremely critical. Water is needed not only to meet the demands for drinking water but to supply industry and farm irrigation systems.

A natural forest cover on slopes and uplands prevents land erosion and the loss of soil nutrients. In this way forests conserve soil fertility. In addition, the forest vegetation increases the rate at which water that reaches the ground surface infiltrates into the ground and it provides an opportunity for groundwater storage; both diminish the impact of flood-producing storms. Forest areas also influence climate. As a protective windbreak, a wooded area can reduce wind velocity by 20 to 60 percent. And a canopy of forest trees can cause cooling of air to temperatures so significantly lower than those in the open that it may be considered a natural air conditioner. In addition, plant cover acts as a cleanser of the air. Plant leaf surfaces filter out ash, dust, and poisonous pollutants and, as mentioned earlier, under some circumstances plant tissues serve as indicators of atmospheric pollution.

Trees and even lower growing shrubs reduce noise pollution by acting as barriers to sound. It is estimated that each 100-foot width of woodland vegetation can absorb about 6 to 8 decibels of sound intensity. Along busy highways where as much as 72 decibels of sound is generated, a forest can give nearby residents much needed relief from noise. This is particularly important to New Jerseyans who, according to state environment officials, live in the noisiest state in the country.

Even the plants of the sand dunes are important to man. The role of vegetation in stabilizing otherwise mobile sand dunes is described in Chapter 13. Illustration of this has been seen in New Jersey as well as in the Netherlands, where a long struggle against encroaching seas has made the people well aware of the importance of maintaining vegetation on the sand dunes. A violent coastal storm in March 1962 caused tremendous damage to coastal and bay areas in New Jersey because storm waves breached some of the ocean-facing dunes. However, in the few places where dunegrass grew abundantly, the sands of the dunes were so stabilized that the seas did not breach the dunes. Houses located thereon remained although others were carried off by the encroaching waters.

Social Values of Natural Vegetation

Less tangible, but still very important, are the social values of natural vegetation. The most obvious of these is the use of natural land areas for recreation.

As generally defined the word "recreation" means a refreshment of one's mind or body through diverting activity. Physicians as well as sociologists now recognize that working or living in an urban locale, a tense atmosphere because of its overcrowding, high levels of noise and atmospheric pollution, littered streets, and impersonal relationships among individuals, contributes to mental and physical illness. For the sake of simple well-being, some type of contrast is needed to offset the urbanized environment with its uninterrupted structures of concrete, brick, glass, and steel, and its hordes of people.

To many individuals a contrast comes only from some form of active outdoor recreation, perhaps swimming, skiing, golfing, horseback riding, boating, fishing, hiking, camping, or hunting. To others, a complete and solitary escape from people is needed — a retreat into a wilderness whether a marsh, bog, or forest. To still others tranquility can come from observation or study of natural life — from bird-watching, amateur botanizing, or a search for fossils or Indian artifacts. And to some, just a short walk in a park, picnicking, or, even, a drive through landscape with scenery different from that of the city will be adequate for refreshment of spirit.

Fulfillment of these varied needs necessitates the preservation of many different tracts of public land. Areas of marshes, bogs, swamps, forested uplands, slopes, ridges, and sand dunes must be available in a size adequate to satisfy not only the diverse needs of the present population but also those of the larger future population.

The Value of Diversity Itself

Much is being written today about the value of diversity. Whether applied to populations of people or to populations of plants, the disadvantages of uniformity are obvious. However, as this book is devoted to a study of landscape diversity, only the value of variety in plant life is considered here.

Certainly, the most pleasant landscape in which to live is one which encompasses a variety of landforms and plant communities—marshes, swamps, and forests of different ages and composition. Such natural diversity, however, may have even greater value than pleasure. Ecologists are not yet completely clear about the relationship between diversity and stability in communities of plants and animals. They know, however, that in the growing of agricultural crops, uniformity in plant culture can be unstable. Extensive areas of land covered with just one crop have been completely devastated either by a single species of insect or by one plant fungus disease. In a mixed growth of plants such as occurs in a natural forest, this normally does not happen. For this reason, many ecologists agree with Dr. Eugene Odum's suggestion that biotic diversity may be more than just "the spice of life." Indeed, it may be a necessity for the continuation of the natural ecosystem for man as well as for other forms of life. Thus, a healthy natural environment is a diverse one and both are synonymous with a healthy environment for man. In this sense, then, preservation of landscape diversity in New Jersey may be of priceless worth.

Major Environmental Issues Facing
New Jersey's Citizens

Currently much is being said and written about environmental problems—a subject which only a few years ago received little public attention. Singly or together, population growth, increased industrialization, modern technology, and current lifestyles are blamed for the degradation of the environment. In a well-publicized report, *The Limits to Growth,* a group of international scientists, educators, economists, humanists, industrialists and public officials have predicted the ultimate consequences of continuing present rates of population increase, industrial growth, use of nonrenewable resources, consumption of food, and environmental pollution—five interrelated factors. The predicted result is not an encouraging one for mankind, but the authors of the report suggest that immediate initiation of constructive actions can prevent the disastrous consequences which otherwise would develop.

We share with others this feeling of urgency about seeking solutions to environmental problems both on a national and on an international front. However, the present discussion is confined to issues that are relevant to New Jersey—specifically, those relevant to the future of the natural vegetation of the state. There are four such issues of immediate and prime concern, all of which transcend the state of New Jersey:

- How much future population growth?
- What future industrial growth?
- How much and what kind of open space?
- Who plans and controls land use?

It is not the intent of the authors to offer arbitrary answers to these questions but rather to suggest the reasons why these particular issues deserve widespread consideration by the citizens and public officials of the state.

*How Much Future Population Growth
in New Jersey?*

Perhaps the most sensitive issue that faces New Jersey (and also the United States as a nation and as an international political unit) is the question of future population growth.

In view of the benefits served by areas of natural vegetation and the fact that increased numbers of people can be accommodated only by encroachment onto these areas, one must wonder how many more people the land of New Jersey can accommodate without causing some bad effects on the present inhabitants. As noted earlier, New Jersey already is the most densely populated state, and the projections of future population growth forecast that the number of people will almost double by the year 2000 (Figure 15-1). Obviously, there is an upper limit of population beyond which additional overcrowding in the state will cause drastic and harmful changes in living conditions.

Unfortunately, scientists do not know what the "optimum" limits of population are for man, but experiments with some animal species have revealed the occurrence of severe psychological disturbances when they are forced to live together under overcrowded conditions. There is no reason to expect that the same would not be true for man, and it is necessary therefore to predict the optimum population figure for man far enough in advance to avoid a disastrous situation. As an initial step and as a function of state planning, the impacts that the forecasted population growth (in numbers and spatial distribution) will have on the environment of New Jersey — specifically on land and water resources, the atmosphere, and natural ecosystems — should be identified. Only when this has been done can the citizens of the state together with public officials truly assess the potential hazards of unlimited population growth. The result may suggest a need for some restriction on the number of residents or on spatial distribution of population in the state.

What Future Industrial Growth in New Jersey?

Industrial growth is interconnected with population growth in several ways. Expansion of service and support industries is needed to accommodate increased numbers of people. Utilities must be expanded to meet demands for electricity, for example. There is continuing pressure also to create greater employment opportunities within the state, and industrial growth is widely hailed as a virtue. State and local public officials as well as private groups, particularly land developers and realtors, are continually trying to lure new industrial establishments to New Jersey.

Increased industrialization involves the exploitation of natural resources. Land is needed for building and for service roads. Huge amounts of water are used by some industries, and industries as a whole are also large consumers of electricity, gas, and other utilities. The disposal of industrial wastes remains one of the major causes of air and water pollution in New Jersey. All this suggests that the present rate of industrial expansion in New Jersey cannot continue indefinitely.

Just as in the case of population, there is an optimum limit to the amount as well as type of industrial growth that can be contained within the state of New Jersey without disastrous effects on the environment. For this reason, it is well that a reorganization of New Jersey state agency functions was effected in April 1970. Before this time, a single agency, the Department of Conservation and Economic Development, had two conflicting responsibilities: first, conserving and protecting environmental resources and, second, promoting industrial expansion in the state. Under the reorganization the single agency was abolished, and responsibility for all functions dealing with conservation and environmental issues were assigned to the newly created Department of Environmental Protection. Responsibilities dealing with planning for economic and industrial growth in the state were transferred to the Department of Labor and Industry.

The new alignment of responsibilities in the state government of New Jersey gives recognition to two conflicting interests:

first, the need to protect natural resources and the environment and second, the desire to have increased industrial and economic growth. The conflict between these two interests must be resolved in the future not only in New Jersey but in the whole of the United States. In the past Americans have been led to believe in the so-called growth myth, which professes that there can be continued and unlimited industrial growth without offsetting disturbances. Only recently has widespread attention been focused on the mixed nature of the effects of growth. It is now more widely recognized that there is an urgent need to reexamine the more or less commonly accepted idea that the good life comes only with continued economic growth. Widespread lip service is being given to the desirability of improving the "quality of life," but as yet there has been no resolution of the conflict between this goal and the desire for unrestrained industrial and economic growth. The conflict is one that must be faced by individual citizens whether students, industrial workers, or educators as well as by the heads of labor organizations, industrial enterprises, educational or scientific institutions, and local, state, and national governmental agencies. The real issue is to what extent and how must economic and technological growth be redirected so as to ensure both a good life for every citizen *and* the preservation of a healthy environment in which the present population and future generations can live safely and happily.

How Much and What Kind of Open Space in New Jersey?

By mid-1972, New Jersey had 12 state forests and 24 state parks; additional land is conserved as national, state, county, local, or private nature preserves. Even so, according to a study made by a state commission, there is not enough open land to accommodate the recreational needs of the existing population, much less those of the projected population. Work weeks have been shortened and vacations lengthened in a trend expected to continue. The increased leisure time of the present population has placed pressures on existing open land facilities. In an official report prepared in 1967 and entitled *New Jersey Open Space Recreation Plan,* it was stated that as of the date of the report, about 158,000 acres of additional public land were needed to meet existing recreation demands in New Jersey. The report added that more than three times this amount of additional public open land will be needed by the year 2000.

A step to acquire some part of the needed land was taken in November 1971 when the voters of New Jersey approved a bond issue of 80 million dollars for the new Green Acres program which permits purchase of additional open space for state, county, and municipal recreation areas. But to acquire enough open land to maintain a healthy environment for the existing and projected population of New Jersey is not the only problem. To conserve the land already acquired requires public support and more stringent legislation. Officially designated parks and wildlife refuges are under continual challenge for other use, and dedicated park land of the state has been confiscated for highways, dams, sewage plants, and public buildings and some has even been transferred to private use. For this reason in 1971 the New Jersey Commission on Open Space Policy recommended legislation requiring that public agencies not take public open space for another use if there is any favorable alternative; and that, if such land must be taken, it be replaced with an equal tract of land in another location.

The need for acquisition of land desirable as public open space is particularly critical in New Jersey because of the limited amount of land relative to the population. These same conditions make it necessary that there be effective management of land acquired for public interest. Some parts of public land must be developed to provide access highways, paved parking areas, and recreational facilities. Additional land must be kept more "natural," so as to fulfill the human desire for areas of greater contrast to the man-created urbanized environment. And in order to preserve for future generations samples of the unique plant and animal life of the state, still more land must be set aside and restricted to such use.

As the population increases, conflicts over the use of public lands will heighten. As in the case of Island Beach State Park, there will be conflicts between those who call for more recreational facilities and those who desire that the land be kept in a natural state. Increasingly the combined efforts of ecologists, foresters, game wardens, sociologists, and experts on recreational facilities will be needed to determine what utilization of lands will best satisfy the diverse public interests. The issue of open space and its use in New Jersey (and the issues of population and industrial growth) obviously cannot be examined and acted upon without consideration for the substantial influences that arise beyond the borders of the state. Among these is the use of the natural recreational areas of New Jersey by the population of surrounding states, particularly New York and Pennsylvania where the largest and fourth largest cities of the United States have a combined population about double that of the entire state of New Jersey.

Who Plans and Controls Land Use in New Jersey?

A fourth issue relates to the responsibility for determining the use that may be made of land. As long as New Jersey, and other states as well, appeared to have an inexhaustible amount of land, there was no overwhelming reason to place restrictions on the rights of individuals to acquire and to develop land without consideration for the effects of land use on the public. The situation has changed and echoing what Aldo Leopold four decades ago urged as a "land ethic," President Nixon stated in a message to Congress on August 10, 1970:

We have treated our land as if it were a limitless resource. Traditionally, Americans have felt that what they do with their own land is their own business. This attitude has been a natural outgrowth of the pioneer spirit. Today, we are coming to realize that our land is finite, while our population is growing. The uses to which our generation puts the land can either expand or severely limit the choices our children will have. The time has come when we must accept the idea that none of us has a right to abuse the land, and that, on the contrary, society as a whole has a legitimate interest in proper land use.

Much has happened just in the two years since the President made this statement. A recent report commissioned by the United States Council on Environmental Quality states, "This country is in the midst of a revolution in the way we regulate the use of our land. It is a peaceful revolution, conducted entirely within the law. It is a quiet revolution, and its supporters include both conservatives and liberals." Evidence of changing attitudes toward land-use controls is seen in New Jersey; these have developed for several reasons.

First, the general public has become aware that the use made of particular types of land or of a particular tract of land may have an impact on many people. For example, filling in a flood-plain in one area may cause severe flooding in another; the use of septic tanks for disposal of sewage for one residential develop-ment may endanger the water supplies of other areas; the dump-ing of raw sewage into streams by one community may pollute the water supplies of other communities and also make recrea-tional use of the stream waters impossible. Compared with many Europeans, Americans have been slow to recognize the "social" value of land – land as a public resource and the need to restrict in the public interest the use that an individual may make of his land. A major step was accomplished when the New Jersey Legis-lature enacted in November 1970 the State Wetland Act which recognized the vital importance of the state estuarine zones and authorized the state Commissioner of Environmental Protection to regulate wetland development – to restrict dredging, filling, re-moving, and polluting of the coastal marshlands. The wetland act was not passed without a struggle. Individuals with private land interests strongly opposed the law, as it prohibits use or development of the land without consideration for the impact on others.

Court actions have also stimulated change in attitudes to-ward land-use controls. The recent court decisions declaring un-constitutional the present method of financing schools by property taxes will have a significant impact on land use. The tax system has exerted great influence over land-use controls because local planning and zoning authorities have been concerned with the need to get "good ratables" and exclude the "poor ratable" uses of land. A different attack on the present land-use control activities has come from court decisions outlawing exclusionary zoning regulations.

These expressions of changing attitudes toward land-use control point up a need for new mechanisms of land-use planning and control. At the same time it has become apparent that local control of land use (in New Jersey by 567 municipalities) is inadequate to cope with the problems that transcend municipality boundaries. This was evident in the attempts of the fourteen municipalities to control individually the parts of the Hackensack Meadows lying within their borders. Recognizing the problem, the New Jersey Legislature in 1968 created the Hackensack Meadowlands Development Commission and charged it with the responsibility to develop an integrated plan for the whole area. Another important attempt at comprehensive regional planning in the state came with the creation of the Pinelands Environmental Council in June 1972; one of the important functions of the council will be preparation of a master plan and guidelines for the use of the Pine Barrens which cover over 373,000 acres in Burlington and Ocean Counties.

Congressional committees are now developing federal legislation relating to land-use control; the pending bills call upon the states to develop comprehensive land-use plans and to ensure the implementation of plans by controlling legislation. The committees recommend that particular consideration be given to areas of critical environmental concern such as beaches, marshes, swamps, and floodplains. If the proposed federal legislation is enacted, New Jersey will have to strengthen its role in land-use planning and control, as will other states. To accomplish this will require an upswelling of public support against private interests who might zealously attempt to stop any encroachment into the present prerogatives of the local municipalities.

The Summary — A Citizen's Responsibility

In the preceding sections some of the critical issues involved in the future relationship between man and his environment in New Jersey have been identified. Preservation of natural vegeta-

Above, solid waste pollution in the Hackensack Meadows. *Photograph by the New Jersey Department of Environmental Protection. On opposite page, upper,* air pollution in Newark industrial area. *Lower,* water pollution in a stream on the Inner Coastal Plain. *Photograph by the New Jersey Department of Environmental Protection.*

Figure 15-3   Pollution in New Jersey.

288   *Vegetation of New Jersey: A Study of Landscape Diversity*

tion is dependent on the solution of other problems too. One of these is pollution. Whether the atmosphere, the streams, or the soils are polluted and whether the source of pollution is industrial wastes, public sewage, the runoff of agricultural fertilizers, the excessive use of toxic pesticides or herbicides, or simply individual acts of vandalism or littering, natural vegetation suffers (Figure 15-3). It is hoped that fears for the survival of man himself will be so widespread that the pollution activities which affect plants as well as man will be curtailed.

The desire to provide a healthy and pleasing environment for the present population and for future generations must be a common goal shared by a majority of citizens in the state. Without a consensus that the goal is sound, the efforts of ecologists, public officials, legislators, and small bodies of concerned citizens cannot be successful in achieving implementation of the policies and programs needed to ensure a safe environment. The first step in gaining wide support for such programs is to acquaint all citizens with the interrelationships between man's actions and the environmental resources on which he depends for survival. Natural vegetation is one of these environmental resources.

To conserve for future generations the diverse natural vegetation that exists in New Jersey today will not be easy. In a state already highly populated and industrialized, a deep-seated desire for endless economic growth conflicts with what Lyndon Johnson stated should be a prime natural goal: "An environment that is pleasing to the senses and healthy to live in."

Actions must be taken now to preserve for present and future generations enough and the right kinds of open space in the right places. But this involves conflicts with individual interests. Only with the support of the majority of the people can the goal of a safe and pleasing environment be achieved.

REFERENCES AND SOURCE MATERIAL

Bosselman, Fred and David Callies. 1971. The Quiet Revolution in Land Use Control. Summary Report. Council on Environmental Quality. U.S. Government Printing Office, Washington, D.C.

Gottmann, Jean. 1961. Megalopolis. MIT Press, Cambridge, Mass.

Johnson, Lyndon B. 1965. Special Message to Congress on Natural Beauty, February 8, 1965.

Leopold, Aldo. 1933. The Conservation Ethic. Journal of Forestry 31: 634–643.

McHarg, Ian L. 1969. Design with Nature. The Natural History Press, Garden City, N.Y.

Meadows, D. H., et al. 1972. The Limits to Growth. Universe Books, New York, New York.

Moore, E. B. 1939. Forest Management in New Jersey. New Jersey Department of Conservation and Development.

New Jersey. 1967. State Department of Conservation and Economic Development. New Jersey Open Space Recreation Plan, Trenton, N.J.

New Jersey. 1969. State Department of Conservation and Economic Development. Bureau of Research and Statistics. 1969 Population Estimate for New Jersey.

New Jersey. 1970. State Department of Environmental Protection. Air, Water and Waste Management Times, Trenton, N.J.

Nixon, Richard B. 1970. Special Message to Congress Sent with First Annual Report of the Council on Environmental Policy, August 10, 1970.

Odum, Eugene P. 1969. The Strategy of Ecosystem Development. Science 164: 262–270.

Wagar, J. Alan. 1970. Growth Versus the Quality of Life. Science 168: 1179–1184.

Appendix I

Where to See Types of Natural Vegetation in New Jersey

The purpose of this section is to offer a guide to those who wish to see actual examples of the natural vegetation types that have been described in this book. For the most part, the sites described herein are in state forests and parks or in other public parks accessible from public roads. The guide material is organized to conform with the sequence of the text chapters (Chapters 7 to 13) which describe the vegetation types.

Salt Marsh Vegetation (Chapter 7)

Salt marshes occur in a number of places along the coast and wide expanses of salt marshland can be seen from the Garden State Parkway south of Exit 58 particularly where the highway crosses the Mullica River and Great Egg Harbor and then southward to Cape May.

One of the best places to see a salt marsh at its outermost edge where one can walk on it and observe it closely is at the Tuckerton marsh in Burlington County. To reach the marsh area take the Great Bay Boulevard that intersects with Highway 9 in Tuckerton; this road goes south of Tuckerton and east of Great Bay and traverses an expanse of marsh terminating at a former Coast Guard station at Great Bay, now used as the Rutgers Marine Science Laboratory.

Cheesequake State Park in Middlesex County also has salt marshes. To visit this marsh one must ask for directions to locate it and get permission at the park headquarters office which is at the park entrance reached from Highway 34 east of the town of Cheesequake. The Cheese-

291

quake area contrasts with the Tuckerton site because it demonstrates the marsh at its inner edge where there is a transition from true salt marsh with cordgrass along the tidal creeks, salt meadow grass behind that and a gradation to freshwater marsh of reed grass and cattails and finally to marginal freshwater swamp forest.

There are small patches of salt marsh grasses in the Sandy Hook natural area; Park naturalists conduct daily tours through this area during the summer. Sandy Hook State Park is in Monmouth County and can be reached from Highway 36. A map available at the Park entrance shows the location of the Nature Center.

Freshwater Marsh Vegetation (Chapter 7)

Freshwater marshes of various sizes occur in many places throughout the state. Small clumps of two of the three dominant plants in our freshwater marshes – cattails and reed grass (Figure 7-4) – may be seen in depressions along roadsides throughout the state where drainage from the highway embankment results in a more or less continuous condition of standing water. Patches of wild rice, the third freshwater marsh dominant, can be seen in moist areas along the roadside in the southern counties of the state, particularly Salem and Cumberland. Larger and more natural freshwater marsh habitats are found in areas that were once covered by glacial lakes and along edges of shallow lakes or slow moving streams.

1. *Cattail Marshes*

The largest freshwater "cattail" marsh now present in New Jersey is in Morris County in the area known as "Troy Meadows," located east of Troy Hills, north and east of Whippany and west of Hanover. While some parts of Troy Meadows have progressed to swamp forest through the process of succession, the cattail community is extensive. On the north Troy Meadows were bounded by Highway 46 but Highways 80 and 280 under construction in mid-1972 will be the new northern boundary of the Meadows. The new highway construction has blocked a former access to the marsh but it can be seen at a distance from the south side of Highway 80 where there is a clearing for high tension power lines. A much closer view of the Troy cattail marsh can be had from Edwards Road which has been rerouted nearer to the east side of the marsh to accommodate the construction of Highway 280. Edwards Road runs to the east of Troy Meadows and can be reached from Highway 46. Where the newly routed portion of Edwards Road crosses Smiths Ditch River just north of the Hanover Airport, the cattail marsh lies close to the west side of the road. Because silting as well as pollution appears to favor reed grass over cattails, it is hoped that this area of marshland can remain in its present form in spite of the disturbance of the highway construction.

Little cattail marshland is accessible for observation in Hatfield Swamp that lies to the east of Troy Meadows, in Great Piece Meadows to the north or in Black Meadows to the south—all four marshes being relicts of Glacial Lake Passaic. But fortunately steps have been taken to preserve as a National Wildlife Refuge the marshes of the Great Swamp which lies south of Black Meadows; it also owes its origin to Glacial Lake Passaic. One may drive through parts of the Great Swamp Refuge and from the road areas of cattail marshes can be seen. A very accessible place to observe the marshes more closely is from the Refuge Wildlife Observation Station located on Long Hill Road, which goes from Meyersville to New Vernon. There is a parking lot at this Wildlife Observation Station and a marked trail leads through a cattail marsh and a swamp woods. Other trails in the Great Swamp Refuge are open to the public and information about these can be obtained from the Refuge Headquarters located on Pleasant Plains Road which can be reached by driving north on Central Avenue from Stirling. The Morris County Park Commission also sponsors trails and guided tours in a part of the Great Swamp. Information about these can be obtained from the Morris County Outdoor Education Center, which is located on Jay Road, reached from Southern Boulevard out of Chatham. A third nature educational facility is located at Lord Stirling Park in Basking Ridge and it is adjacent to the Great Swamp National Wildlife Refuge. This is sponsored by the Somerset County Park Commission and it, too, offers nature educational facilities.

2. *Reed Grass (or Phragmites) Marshes*

The best examples of large tracts of marshland dominated by reed grass are seen in the "Newark Meadows" and "Hackensack Meadows," both on the site of former Glacial Lake Hackensack. All but a small portion of the remaining marshland in these areas is dominated by reed grass which grows in brackish as well as in fresh water. This is the very tall grass with showy seed plumes that one sees in the marshes driving on the New Jersey Turnpike from the Newark Airport north to the George Washington Bridge exits; a good view also can be had when traveling by train between Newark and New York City. The reed grass grows so tall along some of the roads in southern New Jersey that it obstructs view of the salt marshes or wild rice freshwater marshes. Such is the case of the road to Fort Mott Park at the southwestern tip of the state.

3. *Wild Rice Marshes*

Marshland dominated by wild rice, an annual plant (Figure 7-4) conspicuous from late spring to fall can be seen mostly in the southern part of the state. Wild rice is particularly abundant in freshwater marshes along the rivers emptying into the Delaware Bay, but there are good wild rice stands as far north as the Rancocas Creek that flows into the Delaware River. Here a growth of wild rice can be seen from a road that crosses the Creek just west of Rancocas State Park. The road, Bridge Street in Burlington County, runs perpendicular to the road that runs between Timbucton and Willingboro. Driving west from Timbucton to Willingboro, one passes the entrance to Rancocas State Park, and shortly thereafter the road becomes a four-lane dual highway passing over the New Jersey Turnpike and Highway 295. The first left after crossing Highway 295 is Bridge Street marked as the road to Centerton. Adjacent to the left side of this road just before the bridge that spans Rancocas Creek is a stand of wild rice. Crossing the bridge and looking back from the other side of the Creek, stands of wild rice can be seen both to the east and west of the bridge. To observe the wild rice more closely one may park in Rancocas State Park and walk to the Creek. At mid-1972, the park had not yet been developed for recreational use; however, a map of the roads and the site of the natural area in the park has been prepared and may be requested from the New Jersey Department of Forests & Parks, Box 1420, Trenton, New Jersey 08625.

Another accessible place to see wild rice is along the Maurice River in Cumberland County; here it can be viewed from the bridge that spans the river at Mauricetown or one can walk close to the marsh north of Mauricetown where Laurel Lake drains into the river.

Bog Vegetation (Chapter 8)

An excellent example of a mature bog in northern New Jersey that can be visited is in High Point State Park. Cedar bog (locally called Cedar Swamp) lies just northwest of High Point monument and the map of High Point State Park available at the Park entrance identifies the bog area as the "John Dryden Kuser Natural Area." One can drive to within about 100 yards of the bog, park, and then walk along a trail that encircles the bog forest.

Without even getting one's feet wet, it is possible to see clearly the bog vegetation that has completely filled in a lake created upon recession of the glacial ice that covered this part of New Jersey. Along the trail can be seen large trees of hemlock, red maple, and southern white cedar, with some representatives of black gum, yellow birch, and white pine trees and of oaks and sugar maple, the latter two on higher ground. Great banks of rhododendron border some parts of the trail as do clumps of ferns and skunk cabbage; sphagnum moss with herbs typical of bogs grow in the interior of the bog. No open water now remains in this bog but the occasional black spruce and larch trees that occur in the more open bog interior are reminders of the natural succession process in which an open body of water has changed into a site with forest vegetation.

In southern New Jersey there are many places along the edges of streams, particularly in the Pine Barrens, where cedar bogs can be seen. The best views of the various communities of the bog vegetation—the sedge meadows, leatherleaf and blueberry shrub thickets and cedar forests—can be had by canoeing down the rivers (Batsto, Mullica, Oswego, and Wading) that run through the Barrens.

There are several places, however, that cedar bogs are accessible from the roadside; in Lebanon State Forest a road crossing Shinns Branch passes through a cedar bog as does the road that crosses MacDonalds Branch. A map showing these roads can be obtained from the forest headquarters office, which is located off Route 72 just east of the intersection of Routes 70 and 72.

There is a young cedar bog in Bass River State Park in the Lake Absegami area where there is a nature trail going through the bog area. A folder with a map of the trail can be obtained from the forest office at the park entrance which is on Stage Road that goes from Tuckerton to Highway 563.

Cedar bogs can be seen also at Batsto in Wharton State Park, where there is a nature center; guided tours through the natural area are scheduled at regular times in the summer. The entrance to Batsto is on Route 542 east of Hammonton.

Swamp and Floodplain Vegetation in North Jersey (Chapter 9)

A readily accessible place to view an example of a North Jersey swamp forest is in the Great Swamp in Morris County. A roadside view of the typical trees in this forest is available on Pleasant Plains Road, which crosses marshes and swampland. This road can be reached by going north from Stirling on Central Avenue. About .7 mile north of the beginning of Pleasant Plains Road, the road is bordered by a swamp forest with large pin oak, swamp white oak, and red maple trees and occasional large sweetgum and shagbark hickory trees. Along the road spicebush is a common shrub and in summer months skunk cabbage and ferns are conspicuous. Those who want to see more of this type of vegetation should take one of the trails in the Great Swamp Wildlife Refuge, the directions for which are given in the section on "Freshwater Marshes." In early June beautiful clumps of laurel in flower can be seen in these swamp forests.

In the Jockey Hollow area of Morristown National Historical Park, which is south of Morristown, there is another interesting lowland forest to visit. Here one may see large tulip trees which grow abundantly in the area. The best place to view the forest is at the location designated as the "Lunch Area"; here parking space is available and the Primrose Brook Aqueduct Nature Trail has its starting point.

Swamp forests also can be seen on the flood plains of North Jersey. A beautiful flood plain forest is located along the east bank of the Raritan River at the north end of Johnson Park. One can reach it by starting at River Road (Highway 18) at the foot of Hoe's Lane and walking down to the river. This is a typical forested flood plain with a higher well-drained portion near the river and a lower, poorly drained section with fine soils away from the river. The latter supports swamp species, such as pin oak, elm, swamp white oak and willow, while the well-drained part next to the river is a forest of basswood, sugar maple and other trees common also on mesic upland areas in North Jersey. Unfortunately, the future of this beautiful area is not certain. Because of its great interest as natural floodplain vegetation, which is among the finest of New Jersey's natural gardens and its accessibility, it is hoped that either the State of New Jersey or the Middlesex Park Commission will preserve the area as a natural site.

A magnificent growth of flood plain trees can be seen on Bull's Island in the Delaware River which can be reached from Highway 29 about 3.5 miles north of Stockton in Hunterdon County. Here there are huge lowland trees – mostly sycamore and silver maple with some tulip trees, elm, ash, walnut, locust and box elder. The area is owned by the state and parking is available.

The impact of beavers on swamp vegetation can be seen in North Jersey where beavers are active in the streams that flow in the valleys of High Point State Park and Stokes State Forest. In this locale one can walk along the streams and see beaver-constructed dams and the resulting impact on vegetation. One such dam can be reached by walking a short distance from Saw Mill Road, the highway that runs north from Deckertown Turnpike to High Point State Park. Here Big Flatbrook Brook runs parallel and only a few hundred feet to the west of the road. On this stream, about .9 mile north of the Deckertown Turnpike, a sizable dam was built by beavers and later abandoned. Although the dam has been breached, a good part of it still remains and one can even walk on it. The dam caused flooding and the formation of a pond on land which previously was dry enough for tree growth. To construct the dam,

the beavers cut down a number of trees in the area and those that remained were killed by the flooding. As a result, the natural vegetation of
the stream valley underwent a marked change because of the beaver activity.

Swamp Vegetation in South Jersey (Chapter 9)

1. *Inner Coastal Plain Lowlands*

An excellent example of the Inner Coastal Plain Lowland forest occurs on the Assunpink Creek in Mercer County where it borders Monmouth County east of Trenton. A road bisects the swamp forest and
from it can be seen large and abundant sweetgum, red maple and pin
oak trees with some hickory, white oak, black oak, and black gum. This
road can be reached by taking Highway 130 south from Hightstown and
turning left (east) on the road that goes from Windsor to New Sharon.
Follow this road until it crosses the New Jersey Turnpike and shortly
thereafter it intersects on the left (north) Allens Road. From this intersection north the swamp forest borders both sides of Allens Road.

Pigeon Swamp in Middlesex County also has a diversified swamp
forest with large specimens of the swamp trees typical of this type of lowland on the Inner Coastal Plain – the sweetgum, red maple, pin oak and
black gum trees. A good view of the forest and its typical trees can be had
from two roads which cross the swamp area. Pigeon Swamp lies between Highway 130 and the New Jersey Turnpike, east of Deans and
Dayton. It can be reached from Highway 130 by taking Fresh Pond Road
which goes to the right (east) of Highway 130. When driving north on
Highway 130 the turnoff for Fresh Pond Road is just north of the turnoff
marked Dayton-Jamesburg. Fresh Pond Road (about .6 mile north of
Highway 130) crosses a drainage ditch and at this point specimens of
sweetgum with its star-shaped leaves, red maple, pin oak and black gum
border the road. Continue north on Fresh Pond Road to the first intersection which is Dean Rhode Hall Road. Turn right and about .4 mile down
this road is a drainage ditch; on the right side of the road prior to this
ditch is a good view of the sweetgum, red maple, and black gum swamp
forest. Also note that the gray birch grows in open areas along the roadside.

2. *Outer Coastal Plain Swamp Forests*

The Outer Coastal Plain swamp forests of hardwood trees, as described in the text, consist mostly of red maple, black gum and sweetbay
trees. These hardwoods commonly replace cedar bog forests in succes-

sion or after cutting or fire and commonly still have scattered cedar trees growing in them. Examples of this forest can be seen along the streams in Lebanon and Wharton State Forest (see section on South Jersey bogs). In Allaire State Park, southeast of Farmingdale in Monmouth County, there is a nature trail (the red trail) that crosses a swamp forest. From the boardwalk one can see large specimens of red maple and black gum and some smaller sweetbay. Allaire State Park is reached from a marked exit on the Garden State Parkway.

Parvin State Park west of Vineland in Salem County reached from Highway 540 has typical swamp forest along its Muddy Run area and in Belleplain State Forest in Cape May County the swamp hardwoods forest (and cedar bog) can be seen from the Beaver Causeway Road. In both parks maps showing these locations can be obtained from the park headquarters office.

A good example of the Cape May lowland forest can be seen in Timber and Beaver Swamp which lies southeast of South Dennis. The area can be reached by taking a small road that goes to the east off Route 585 about a mile south of South Dennis. The trees present in this forest are, in order of decreasing dominance, holly, red maple, sweetgum, black gum, loblolly pine, southern red oak, sweetbay and willow oak. The gently undulating terrain makes possible this diverse composition of lowland trees. The loblolly pine is particularly interesting since it is at or close to its northern limit here.

Upland Forests of North Jersey (Chapter 10)

1. *Mixed Oak Forest*

New Jersey has an exceptionally fine example of a Mixed Oak forest that dates back to Indian times. It is the Hutcheson Memorial Forest in Somerset County in the Piedmont section of the state. The three oaks typical of the Mixed Oak forest type in North Jersey — the white, black, and red oaks — are the dominant trees; the associate members of the community are described in Chapter 10. The forest is located west of New Brunswick and just east of East Millstone. Its entrance is on the south side of Amwell Road (Route 514) and is identified by a memorial archway and a sign.

The Hutcheson Memorial Forest has been preserved thanks to the United Brotherhood of Carpenters and Joiners of America who bought the 65-acre forest and gave it to Rutgers University to administer. In addition, a substantial endowment, established from funds contributed by many people initially for purchase of the forest, has made possible research in the forest which has enhanced the value of the conducted tours. The forest is open to the public at scheduled times when tours are conducted. A schedule of the regular tours can be obtained from the Director of the Hutcheson Memorial Forest, Rutgers University, New Brunswick, New Jersey.

In the Watchung Reservation in Union County the Mixed Oak forest with white and black oak particularly abundant can be seen on the hill above Glenside Avenue in the area known as "Deserted Village." In the Herrontown Woods, off Snowden Lane in Princeton, in Mercer County, large red oaks as well as white and black oaks dominate the upland diabase slopes. A booklet describing this natural preserve can be obtained from the Stony Brook-Millstone Watersheds Association (see reference in Chapter 10).

2. *Hemlock–Mixed Hardwoods Forest*

The hemlock-dominated forests are usually relatively small in areal extent, but occur frequently in the ravines of northern New Jersey and in the Watchung and Palisades areas of the Piedmont. One of the most readily accessible hemlock stands is located in the Tillman Ravine in Stokes State Forest, Sussex County. This ravine borders a road which runs from Wallpack Center through Stokes State Forest to Route 206; there are two parking areas on this road marked Tillman Falls Recreation Area and Tillman Falls Scenic Area. From both places one can pass through a good example of the hemlock-dominated forest when walking downslope to the stream in the ravine.

As typical of this type of forest, almost all the trees in the forest are hemlock including the mature trees and developing seedlings and saplings. In this forest some of the trees, over 100 feet in height and almost four feet in diameter, are thought to be over 150 years in age. In openings, a sassafras or sprouts of a chestnut tree may be seen. Large clumps of rhododendron occur in this forest and there are occasional representatives of witch hazel, laurel and blueberry shrubs. The spicebush may be seen along the bank of Tillman Brook. The forest floor, as typical in hemlock-dominated woodlands, is covered with hemlock needles and there is little plant ground cover; partridge berry is the only herb of importance although in spring trailing arbutus and several herbs can be seen.

A note of caution about plant identification here: the hemlock tree should be distinguished from nearby red pines and white pines, both of which are evergreen conifer trees but with much longer needles. The red pine, rare as a native tree in New Jersey, was planted in large sections in the area and plantations of it are seen from the road that passes through the Tillman Ravine area. The red pine is distinguished from the native white pine by having two needles rather than five in a single bundle. A map showing the location of the Tillman ravine and the trails in the area can be obtained from the Stokes State Forest office located on a road just off Route 206.

A hemlock-dominated forest also can be seen across the stream from the "Deserted Village" in the Watchung Reservation in Union County.

3. *Sugar Maple–Mixed Hardwoods Forest*

New Jersey still has examples of the mature Sugar Maple–Mixed Hardwoods forest type. Diverse in numbers of tree species and undergrowth, the forests are beautiful in all seasons and especially so in spring when a wealth of wild flowers cover the forest floor before the large canopy trees are in full foliage. Unfortunately, some of the more mature sugar maple–dominated forests are not easily accessible, but many of the valleys in the Ridge and Valley and Highland sections in northwestern New Jersey still have roadside woodlands in which sugar maple and a good mixture of other hardwood trees grow.

One convenient place to see a Sugar Maple–Mixed Hardwoods forest is at High Point State Park in Sussex County. Deckertown Pike, the road that goes from Libertyville to Montague, crosses through the southern part of High Point Park, and in one valley the road borders a Sugar Maple–Mixed Hardwoods forest. The forest can be entered by foot from the south side of Deckertown Pike at a point of about 1.5 miles west of Saw Mill Road, the road leading to High Point monument. In addition to sizable sugar maple trees, this forest has a good representation of other hardwoods—the yellow birch, sweet birch, basswood, red oak, white oak, red maple, and ash; the beech tree is common in places. Dogwood, ironwood and hop hornbeam are the lower growing trees in the forest and a few chestnut sprouts can be seen. The common shrubs are witch hazel and viburnum with spicebush abundant along the stream bank. Young sugar maple seedlings are particularly numerous in the forest and although the forest floor is rocky there is a rich herb layer, species of ferns are especially common.

Successional Vegetation Types of North Jersey Uplands

Examples of the successional stages of vegetation that follow on sites in the Piedmont section where agriculture has been abandoned can be seen on the property of the Hutcheson Memorial Forest; the stages are pointed out on the guided tours mentioned above in connection with the "Mixed Oak Forest." Here, as throughout the Piedmont, the red cedar is the conspicuous tree of the early woodland stages of succession.

North of the Piedmont, in areas of the Highlands and the Ridge and

Valley Sections, the plant species in the successional stages of old fields differ from those of the Piedmont. At Voorhees State Park, which is just north of the town of High Bridge and can be èntered from Highway 513, one can find typical North Jersey early woodland stages dominated by gray birch and large-toothed aspen. These same two trees can be seen in many places at the edges of the forests in High Point State Park and Stokes State Forest. For example, Highway 23, which goes from Coles-ville to Port Jervis, just before the entrance to High Point on the north side of the road, clumps of gray birch and aspen have developed and dominate the outer edge of the woodland.

Forests of the North Jersey Ridges (Chapter 11)

1. *Chestnut Oak Forest*

Typical ridgetop Chestnut Oak forests can be seen along the Appa-lachian trail that follows the Kittatinny Ridge in High Point State Park and Stokes State Forest. One of the most accessible locations to visit an unusual example of this forest is that part of the Appalachian trail that crosses Deckertown Pike, the road that runs from Libertyville to Montague in Sussex County. The forest is located on the Deckertown Turnpike about .9 miles west of Brink Road and 1.3 miles east of the Saw Mill Road that goes north to the High Point monument. Another accessible place to view the forest is from the parking area at the Sun-rise Mountain outlook which is in Stokes State Forest.

The winter climate is severe in the ridgetop location and tree tops are broken off by ice storms and high winds. The height of the forest growth is thus restricted — a feature characteristic of the ridgetop vege-tation of northern New Jersey. In these forests the chestnut oak tree is by far the most abundant tree but one may see some representatives of red maple, black oak, red oak, scarlet oak, white oak, hickory, and sweet birch in the forest. Chestnut tree sprouts are common but these grow only about 10 to 15 feet in height before dying back. The heath shrubs are common, particularly the blueberries and huckleberries, but herbs other than ferns are sparse in numbers.

2. *Pitch Pine–Scrub Oak Forest*

An example of the Pitch Pine–Scrub Oak forest of North Jersey can be seen when driving along the Scenic Drive in High Point State Park. A map showing the location of this road can be obtained at the entrance gate to the Park. A good place to see the Pitch Pine–Scrub Oak forest is on the exposed ridge on the north side of the Scenic Drive where one can overlook Lake Marcia. Here can be seen stunted representatives of pitch pine trees with an understory of scrub oak and the characteristic heath shrubs, particularly huckleberry and blueberry, the vegetation typical of areas in the Pine Barrens in South Jersey. In High Point this type of vegetation is growing on thin, rocky soil rather than on deep sand as in the Pine Barrens. The plant growth is similar in the two areas because of the chemical quality of the soil (in both locations the soils are acid with low fertility) and because of recurrent fires that are a controlling factor of plant development in both environments.

Vegetation Successional Stages on Rocks (Chapter 11)

In northern New Jersey the glacial ice sheets scraped away from rock surfaces the soil cover as well as the existing vegetation. Now various stages of successional vegetation appear on such rocks. A good location to examine the development of vegetation over rock surfaces is on outcroppings along the Kittatinny Ridge. Rock outcroppings are common in the area surrounding the monument in High Point State Park and on these various stages of rock successional vegetation can be seen such as the growth of lichens that look like stains on rocks. In cracks in the rocks, hairgrass and other herbs grow, and where some soil has accumulated a shrub such as a blueberry, huckleberry, or laurel may be developing or even a tree seedling of the scrub oak, pitch pine, chestnut oak, sweet birch or white birch. In more moist areas of the park, such as in the woodland around the parking area near Cedar Swamp (the John Dryden Kuser Natural Area), these are rock outcroppings with moss mats on which shrubs or tree seedlings may be developing.

Mixed Oak and Beech Oak Forests of the South Jersey Mesic Uplands (Chapter 12)

A very interesting example of the South Jersey Uplands Mixed Oak forest occurs in the upland areas of Rancocas State Park in Burlington County. (Directions for the location of this park are given in the section on freshwater marshes.) After entering the park, take the first road to the right and along this road the forest has as its tree components: black oak and white oak with a few chestnut oak, red maple, tulip tree, sweetgum and some American holly trees—the last more typically found in sand dune woodlands and in Mixed Oak forests along the coast. Sprouts of the chestnut tree are abundant and the heaths, particularly blueberry, huckleberry, and laurel form the shrub cover.

Another good example of the Mixed Oak forest of South Jersey can be seen at the entrance of Allaire State Park in Monmouth County. The Park is located south of Farmingdale and can be reached from a marked exit on the Garden State Parkway. On the right of the road just at, the entrance of the park is a fine Mixed Oak forest with a mixture of white oak, black oak, and chestnut oak; holly trees can be found here and mountain laurel in this locale is abundant in the shrub layer.

The uplands of Cheesequake State Park (see section on Marshes) also are covered with the Mixed Oak forest type with white oak, black oak, and chestnut oak very abundant. Red maple and hickory also are common and laurel, blueberry, and huckleberry are the most abundant shrubs. In this park area gray birch is common along the roadside.

Only remnants of the Beech Oak forests are left on the highly developed mesic uplands of the Inner Coastal Plains. These are private woodlands and most are not accessible. One may view from the roadside good growth of mature beech trees in the Pigeon Swamp area (see section on South Jersey Forests—the Inner Coastal Plain Lowlands). On the left side of Fresh Pond Road about .3 mile north of the Dean Rhode Hill intersection is a small grove of large beech trees. Another beech woodland can be seen from the roadside west of Chesterfield in Burlington County. On the right side of the road from Chesterfield to Arneytown, about 1.7 miles east of the intersection where Highway 528 turns south to Jacobstown, is a privately owned forest in which the beech tree is abundant with some representatives of the white oak, ash, and tulip trees.

Successional Forests of the South Jersey Mesic Uplands (Chapter 12)

Sweetgum Successional Forest

One type of successional forest found on the Inner Coastal Plain is dominated by the sweetgum tree. One such stand is located in Burlington County just northwest of New Lisbon. It can be seen from New Lisbon Road which intersects Highway 530 that goes from Pemberton to Browns Mills. On the right side of New Lisbon Road going toward New Lisbon, about .4 to .5 mile from the intersection with Highway 530, is a good example of successional sweetgum forest.

Another fine example of a sweetgum successional forest can be seen from a nature trail at Allaire State Park (for directions see section on Mixed Oak Forest of South Jersey). On the red trail in this Park just before reaching the area identified as Brickfield, there is an even-aged sweetgum woodland.

Virginia Pine Successional Forest

As noted in the text, stands of the Virginia pine tree are limited to particular locales in New Jersey but one such forest can be seen in Ocean County at the intersection of Highway 539 and the road that goes from New Egypt to Colliers Mills. The forest is on an embankment at the northeast corner of the intersection. As one may observe here, the appearance of the Virginia pine tree is quite different from that of the pitch pine, the characteristic tree of the Pine Barrens. The needles of Virginia pine grow two in a cluster and are shorter than those of the pitch pine. Also, each needle leaf is twisted giving the tree a unique appearance.

Forests of the Drier Pine Barrens (Chapter 12)

1. *The Barrens Oak-dominated Forest*

The Oak-dominated forest can be seen from the roadside in many places in the Barrens. An accessible place to park and observe the forest is in Lebanon State Forest which is in Burlington and Ocean Counties. At the traffic circle intersection of Highways 70 and 72 a road goes north to New Lisbon through the State Forest. The first road to the right off the New Lisbon Road is marked Deep Hollow Road and an Oak-dominated forest type borders both sides of the roadside to Deep Hollow. Controlled burning as described in the text has been practiced in this woods as a fire prevention measure and there is little litter accumulation on the forest floor. The black, white and chestnut oaks are the most common large trees along the road with a few pitch pine. One can see here examples of black and chestnut oak trees with multiple sprouts. The oak forest without controlled burning can be seen in Lebanon State Forest on Shinns Road going to Pakim Pond. A map showing these roads in the forest can be obtained from the Lebanon Park Office which has its entrance on Highway 72, east of its intersection with Highway 70.

2. *The Barrens Pine-dominated Forest*

One passes through extensive trails of the typical Pine-dominated forest when traveling through the Barrens. But again the State Forests offer the most accessible places to view the forests more closely. In Lebanon Forest pitch pine dominates the forest bordering Cooper Road that runs from Pakim Pond to the "200 Area" and on the roads going north from here, such as Glass Works Road.

Driving south on Highway 539 about 5 miles south of Warren Grove (below the area of the Pine Plains) one passes through large areas of the Pine-dominated forests that were badly burned by fire in recent years. Note here that some of the large pines are able to put out new needles while others have died off.

3. *The Barrens Pine Plains (or Dwarf-Pine Community)*

Two highways pass through the areas of the Pine Plains; Highway 72 crosses the West Plains and provides a good place to stop to see the vegetation of the Pine Plains. Traveling east on Highway 72 from its intersection with Highway 70, one reaches the highest point of land about 3.5 miles before reaching the intersection with Highway 539. Here on the south side of the Highway one can park and walk into the Pine Plains which extend to the east and south. Another section of the Pine Plains (the East Plains) borders the west side of Highway 539 about 2 miles south of Warren Grove.

Sand Dunes Vegetation (Chapter 13)

The distinctive, unspoiled vegetation of the New Jersey sand dunes can best be seen at Island Beach State Park which can be reached from Highway 37 east of the Garden State Parkway Exit 82. One can see the types of sand dune plant communities described in the text from a distance as one drives down the main roadway through the park but a better way to get to know the dunes and their vegetation is to arrange to take a guided tour with the park naturalist. This can be arranged at the park nature center where information is also available about the nature trail.

At Sandy Hook State Park in Monmouth County there also is a nature center, trails, and guided tours. The Park is reached from Highway 36 and a map which can be obtained at the entrance gate shows the location of the nature center. On the guided tours the famous holly trees of the dune forest can be seen.

Appendix II

References for Plant Identification

Trees

Harlow, William M. 1957. Trees of Eastern and Central United States and Canada. Dover Publications, Inc. New York. 288 pages. Paperback, 4½″ x 6⅜″.

Designed by size for field use, this book includes descriptions of the features and range of 140 common tree species that grow in Eastern United States and Canada and also has illustrations of leaves, bark, twigs, and fruit of each tree.

Harlow, William M., and Ellwood S. Harrar. 1958. Textbook of Dendrology. McGraw-Hill Book Co., New York. 512 pages. Hard cover, 6¼″ x 9¼″.

This is one of the classic texts for identifying the botanical features and ranges for 150 tree species of greatest importance to forestry in North America. The book is richly illustrated with photographs and tables that highlight distinguishing tree characteristics.

Trees and Shrubs

Harlow, William M. 1959. Fruit Key and Twig Key to Trees and Shrubs. Dover Publications, Inc. New York. 56 Pages. Paperback, 5⅜″ x 8½″.

This is a handy field guide for winter identification of woody plants. It provides keys and descriptions for both fruit and twig features of the common woody plants of Eastern North America.

Petrides, George A. 1958. A Field Guide to Trees and Shrubs. Houghton Mifflin Co. Boston. 431 Pages. Hard cover, 5″ x 7½″.

This book is designed as a field guide for the identification of 645 species of shrubs and trees that grow in Northeastern and North Central North America. The identifying features of each species are described and the leaves, twigs, and barks of some are illustrated. The arrangement of species descriptions is by groupings of plant structural features.

Trees, Shrubs, and Herbs

Gleason, H. A. 1962. Plants of the Vicinity of New York. Hafner Publishing Co. New York. 307 pages. Hard cover, 5½″ x 7½″.

This book, written by an eminent botanist at the New York Botanical Gardens was designed to aid the identification of the more common wild plants that grow within two hundred miles of New York City. The circumscribed area includes northern and central New Jersey as well as the Inner Coastal Plain and the Pine Barrens region in South Jersey. The book is not extensively illustrated but good keys are given to plant structural characteristics.

Herbs – Wild Flowers

Peterson, Roger Tory, and Margaret McKenny. 1968. A Field Guide to Wildflowers. Houghton Mifflin Co., Boston. 420 pages. Hard cover, 5″ x 7½″.

This book contains descriptions of almost 1,300 species of herbs and a few showy flowering shrubs and woody vines that grow in Northeastern and North Central North America. Flowers of many species are illustrated, some in color. Ranges, habitats, and flowering periods are given for each plant. To facilitate identification in the field, the plant descriptions are classified by color of plant flowers.

Rickett, Harold. 1965. Flowers of the United States: The Northeastern States. Vols. I and II. McGraw-Hill Book Co., New York. Vol. I, 559 pages; Vol. II, 280 pages. Hard cover, 10″ x 13″.

By its sheer size, this is not a publication to carry into the field for identification *in situ*. But developed under the sponsorship of the New York Botanical Gardens, these volumes are recognized as the most comprehensive guide to wild flowers of the United States. Each species is beautifully illustrated and accompanied by a scientifically accurate yet nontechnical description.

Herbs — Ferns

Chrysler, M. A., and J. E. Edwards. 1947. The Ferns of New Jersey. Rutgers University Press, New Brunswick, N.J. 201 pages. Hard cover, 6″ x 9″.

This book includes keys, descriptions and maps showing the distribution of specific ferns in New Jersey. Each species is illustrated by one or more photographs showing the plant and in many cases the habitat. This book is now out of print but is available in many libraries.

Wherry, Edgar T. 1961. The Fern Guide. Doubleday & Co., Inc. Garden City, New York. 318 pages. Hard cover, 4⅝″ x 7½″.

This book contains descriptions of 135 species of ferns that grow in Northeastern and Central United States and adjacent Canada. The features, range, habit, and culture of each fern are described and the characteristic plant parts illustrated.

Herbs — Grasses

Hitchock, A. S. 1950. Manual of the Grasses of the United States. U.S. Government Printing Office. Washington, D.C. 1051 pages. Hard cover, 6″ x 9″.

This is the classic text for identification of grasses. It provides a complete description and illustration of the grasses growing naturally in the United States.

Pohl, Richard W. How to Know the Grasses. Wm. C. Brown Co. Publishers, Dubuque, Iowa. 244 pages. Soft cover, spiral binding, 5½″ x 8″.

Keys and illustrations are given for identification of 326 species of grass that grow in North America.

Other References

Conard, H. S. 1956. How to Know the Mosses and Liverworts. Wm. C. Brown Co. Publishers, Dubuque, Iowa. 226 pages. Soft cover, spiral binding, 5½″ x 8″.

This book provides a good introduction to the study of mosses and liverworts. It starts with a description of the life cycle, classification and characteristics of mosses and liverworts and then provides keys to enable one to identify the names of mosses and liverworts that he may find.

Hale, Mason E. 1969. How to Know the Lichens. Wm. C. Brown Co. Publishers, Dubuque, Iowa. 226 pages. Soft cover, spiral binding, 5½" x 8".

Keys and illustrations are given for common lichens found in North America.

Smith, Alexander H. 1958. The Mushroom Hunter's Field Guide. University of Michigan Press, Ann Arbor. 195 pages. Hard cover, 5½" x 11".

This is a good introductory field guide for the identification of over 120 kinds of mushrooms that grow in Northeastern, North Central and Western United States. Well illustrated, the text describes the important field characteristics of each species, where and when to find it, and comments on the edibility of the plant.

Appendix III

Plant Names

Appendix III contains a two-way cross reference between the common and scientific plant names for all plants mentioned in the text.

The scientific name for any particular plant consists of two Latin words, the first classifies the plant by its genus category (like our surname) and the second identifies the plant within a genus by its species name (like our first name). Simply defined, a plant species is a particular type of plant that maintains its identity because it generally does not interbreed with other plant species.

Some plants, usually only those which are very rare, do not have common names and for this reason are referred to only by their scientific names. On the other hand, some widely distributed plants are known by more than one common name. For example, the tulip tree is also called the yellow poplar tree. Wherever more than one common name is frequently used when referring to a particular plant in New Jersey, the alternate name is shown in parenthesis—"tulip tree (or yellow poplar)."

In some cases a widely used common name is not specific and refers to a number of species within a particular genus. For example, the common name "dogwood" refers to several different types of shrubs as well as the flowering dogwood tree. Wherever a common name does refer to more than one species, the corresponding scientific name is listed with the genus name followed by the abbreviation "spp." which means that it refers to more than one species. For example, the common plant name "hickory" is listed with a scientific name of "Carya spp." indicating that in the genus "Carya" there is more than one species known as hickory.

Botanists do not always agree on the scientific names to be used for identification of plants. The scientific names used herein conform to the nomenclature given in *Gray's Manual of Botany*.

To facilitate reference, a two-way cross reference between the common and scientific names follows:

Appendix III, Part A, lists the common names in alphabetical order giving the scientific name for each plant.

Appendix III, Part B, lists the scientific names in alphabetical order giving the corresponding common names.

APPENDIX III REFERENCES

Bailey, L. H. 1963. How Plants Get Their Names. Dover Publications, Inc. New York. 181 pages.

Fernald, Merritt L. 1950. *Gray's Manual of Botany*. American Book Company, New York.

Appendix III, Part A—
Cross Reference, Plant Common Names
to Plant Scientific Names

Common Name	Genus and Species Name
Alder, speckled	Alnus rugosa
Alder, black (or Winterberry)	Ilex verticillata
Alder, common	Alnus serrulata
Anemone, rue	Anemonella thalictroides
Apple	Pyrus malus
Arethusa	Arethusa bulbosa
Arrow-arum	Peltandra virginica
Arrow-head	Sagittaria latifolia
Arrowwood	Viburnum dentatum
Ash	Fraxinus spp.
Green ash	Fraxinus pennsylvanica
White ash	Fraxinus americana
Aspen (or Poplar)	Populus spp.
Large-toothed aspen	Populus grandidentata
Trembling (or Quaking aspen)	Populus tremuloides
Asphodel, bog	Narthecium americanum
Aster	Aster spp.
Bog aster	Aster nemoralis
Stiff-leaved aster	Aster linariifolius
White-topped aster	Sericocarpus asteroides
Woodland aster	Aster divaricatus
Azalea, swamp (or Swamp honeysuckle)	Rhododendron viscosum
Baneberry	Actaea spp.
Barberry	Berberis spp.
Basswood	Tilia americana
Bayberry	Myrica pensylvanica

Common Name	Genus and Species Name
Beachgrass (Dunegrass or Marram)	Ammophila breviligulata
Beach heather (or Heath)	Hudsonia tomentosa
Beach pea	Lathyrus maritimus
Beach plum	Prunus maritima
Bearberry	Arctostaphylos uva-ursi
Beardgrass	Andropogon spp.
Beech	Fagus grandifolia
Beech drops	Epifagus virginiana
Bellwort	Uvularia spp.
Bent grass	Agrostis spp.
Birch	Betula spp.
Gray birch	Betula populifolia
Paper (or White) birch	Betula papyrifera
River birch	Betula nigra
Sweet (or Black) birch	Betula lenta
Yellow birch	Betula lutea
Bittersweet, climbing	Celastrus scandens
Black alder	Ilex verticillata
Black-eyed Susan	Rudbeckia serotina
Black (or Black-marsh) grass	Juncus gerardi
Black haw	Viburnum prunifolium
Black marsh (or Black) grass	Juncus gerardi
Black oat grass	Stipa avenacea
Blackberry	Rubus allegheniensis
Bladderwort	Utricularia spp.
Blazing star	Liatris graminifolia
Blue flag	Iris versicolor
Blue grass	Poa spp.
Bluebells (or Virginia cowslip)	Mertensia virginica
Blueberry	Vaccinium spp.
Black highbush blueberry	Vaccinium atrococcum
Highbush blueberry	Vaccinium corymbosum
Lowbush blueberry	Vaccinium vacillans
Bluestem grass, little	Andropogon scoparius
Bog rosemary	Andromeda glaucophylla
Boneset	Eupatorium perfoliatum
Box elder (or Ash-leaved maple)	Acer negundo
Bracken fern	Pteridium aquilinum
British soldier	Cladonia cristatella
Brome grass	Bromus spp.
Broom crowberry	Corema conradii

Common Name	Genus and Species Name
Broom sedge	Andropogon virginicus
Brownie's buttons	Biatorella clavis
Bulrush	Scirpus spp.
Bur reed	Sparganium spp.
Bur sage	Franseria spp.
Bush clover	Lespedeza spp.
Butter-and-eggs	Linaria vulgaris
Buttonbush	Cephalanthus occidentalis
Cactus plants	Cactaceae family
California laurel	Umbellularia californica
Cardinal flower	Lobelia cardinalis
Catbrier (or Greenbrier)	Smilax spp.
Cattail	Typha spp.
Cedar, red	Juniperus virginiana
Cedar, Southern white	Chamaecyparis thyoides
Celery, wild	Vallisneria americana
Chain fern	Woodwardia spp.
Chamiso	Adenostoma fasciculatum
Cherry	Prunus spp.
Black cherry	Prunus serotina
Sweet cherry	Prunus avium
Chestnut, American	Castanea dentata
Chickweed	Stellaria spp.
Chicory	Cichorium intybus
Chinquapin, giant	Castanopsis chrysophylla
Chokeberry, black	Pyrus melanocarpa
Chokeberry, red	Pyrus arbutifolia
Christmas fern	Polystichum acrostichoides
Cinnamon fern	Osmunda cinnamomea
Cinquefoil	Potentilla spp.
Climbing fern	Lygodium palmatum
Club moss	Lycopodium spp.
Club moss, Carolinian	Lycopodium carolinianum
Columbine	Aquilegia canadensis
Corn chamomile	Anthemis arvensis
Cotton grass, tawny	Eriophorum virginicum
Cow-wheat	Melampyrum lineare
Crabgrass	Digitaria spp.
Cranberry	Vaccinium macrocarpon
Cranesbill	Geranium maculatum
Creosote bush	Larrea spp.

Common Name	Genus and Species Name
Crowberry, broom	Corema conradii
Curly-grass fern	Schizaea pusilla
Daisy, white (or Oxeye daisy)	Chrysanthemum leucanthemum
Dandelion	Taraxacum officinale
Dangleberry (or Blue huckle-berry)	Gaylussacia frondosa
Deerberry	Vaccinium stamineum
Dewberry	Rubus flagellaris
Dogwood	Cornus spp.
Flowering dogwood	Cornus florida
Gray dogwood	Cornus racemosa
Pagoda (or Alternate-leaved) dogwood	Cornus alternifolia
Red dogwood	Cornus stolonifera
Silky dogwood	Cornus amomum
Duckweed	Lemna minor
Dunegrass (or Beachgrass)	Ammophila breviligulata
Dusty miller	Artemesia stelleriana
Dutchman's breeches	Dicentra cucullaria
Eelgrass	Zostera mariana
Elderberry	Sambucus canadensis
Elm	Ulmus spp.
American elm	Ulmus americana
Red (or Slippery) elm	Ulmus rubra
Enchanter's nightshade	Circaea spp.
Evening-primrose	Oenothera biennis
Fanwort	Cabomba caroliniana
Fetterbush	Leucothoe racemosa
Fir	Abies spp.
Foxtail grass	Setaria spp.
Frostweed	Helianthemum canadense
Gallberry	Ilex glabra
Garlic, wild	Allium canadense
Gentian	Gentiana spp.
Pine barrow gentian	Gentiana autumnalis
Ginger, wild	Asarum canadense
Glasswort	Salicornia spp.
Goat's-rue	Tephrosia virginiana

Common Name	Genus and Species Name
Golden-club	Orontium aquaticum
Golden-crest	Lophiola americana
Goldenrod	Solidago spp.
Blue-stemmed goldenrod	Solidago caesia
Seaside goldenrod	Solidago sempervirens
White goldenrod	Solidago bicolor
Gooseberry	Ribes spp.
Grape	Vitis spp.
New England grape	Vitis novae-angliae
Summer grape	Vitis aestivalis
Grass-pink	Calapogon pulchellus
Greenbrier (or Catbrier)	Smilax spp.
Ground cedar	Lycopodium complanatum
Ground pine	Lycopodium obscurum
Gum, black (or Sour gum)	Nyssa sylvatica
Gum, sweet	Liquidambar styraciflua
Gumbo limbo	Bursera spp.
Hackberry	Celtis occidentalis
Haircap moss	Polytrichum juniperinum
Hairgrass	Deschampsia flexuosa
Hairy-vetch	Vicia villosa
Hawkweed	Hieracium spp.
Hawthorn	Crataegus spp.
Hazel (or Hazel nut)	Corylus spp.
American hazel	Corylus americana
Beaked hazel	Corylus cornuta
Heath plants (for plant names see footnote at end of listing)	Members of the Ericaceae family
Heather, beach (or Beach heath)	Hudsonia tomentosa
Heather, flase (or Golden heather)	Hudsonia ericoides
Hemlock	Tsuga canadensis
Hepatica	Hepatica americana
Hickory	Carya spp.
Red hickory	Carya ovalis
Mockernut	Carya tomentosa
Pignut	Carya glabra
Shagbark hickory	Carya ovata
Hog peanut	Amphicarpa spp.
Holly, American	Ilex opaca
Honeysuckle	Lonicera spp.
Japanese honeysuckle	Lonicera japonica

Common Name	Genus and Species Name
Hop hornbeam	Ostrya virginiana
Hornbeam, American (or Iron-wood)	Carpinus caroliniana
Horse-weed	Erigeron canadensis
Huckleberry	Gaylussacia spp.
Black huckleberry	Gaylussacia baccata
Blue huckleberry (or Dangle-berry)	Gaylussacia frondosa
Impoverished grass	Panicum depauperatum
Indian grass	Sorghastrum nutans
Indigo, wild	Baptisia tinctoria
Inkberry	Ilex glabra
Ipecac, wild	Euphorbia ipecacuanhae
Ironwood (or American Hornbeam)	Carpinus caroliniana
Jack-in-the-pulpit	Arisaema triphyllum
Jewelweed (or Touch-me-not)	Impatiens spp.
Joe-Pye-weed	Eupatorium purpureum
Juneberry (or Shadbush or Service berry)	Amelanchier spp.
King Devil (Hawkweed)	Hieracium pratense
Labrador tea	Ledum groenlandicum
Lady's slipper	Cypripedium spp.
Stemless lady's slipper	Cypripedium acaule
Larch, American (or Tamarack)	Larix laricina
Laurel	Kalmia spp.
Mountain laurel	Kalmia latifolia
Sheep laurel	Kalmia angustifolia
Leatherleaf	Chamaedaphne calyculata
Lily-of-the-valley, false	Maianthemum canadense
Lobelia	Lobelia spp.
Locust, black	Robinia pseudo-acacia
Loosestrife, swamp (or Water wil-low)	Decodon verticillatus
Loosestrife, whorled	Lysimachia quadrifolia
Madrone	Arbutus menziesii
Magnolia (or Sweetbay)	Magnolia virginiana
Maleberry	Lyonia ligustrina

Common Name	Genus and Species Name
Mangrove, red	Rhizophora mangle
Manzanita	Arctostaphylos tomentosa
Maple	Acer spp.
Ash-leaved maple (or Box elder)	Acer negundo
Norway maple	Acer platanoides
Red maple	Acer rubrum
Red maple, three-lobed variety	Acer rubrum var. trilobum
Silver maple	Acer saccharinum
Striped maple	Acer pensylvanicum
Sugar maple	Acer saccharum
Marginal shield fern	Dryopteris marginalis
Marram grass (or Dunegrass)	Ammophila breviligulata
Marsh elder	Iva frutescens
Marsh fern	Dryopteris thelypteris
Marsh fleabane	Pluchea spp.
Marsh mallow (or Mallow rose)	Hibiscus palustris
Marsh rose	Rosa palustris
May apple	Podophyllum peltatum
Milfoil (or Yarrow)	Achillea millefolium
Milkweed	Asclepias spp.
Milkwort	Polygala spp.
Milkwort, orange	Polygala lutea
Mistletoe, American	Phoradendron flavescens
Mockernut (a Hickory)	Carya tomentosa
Moonseed	Menispermum canadense
Mountain laurel	Kalmia latifolia
Mouse-eared chickweed	Cerastium vulgatum
Mullein, common	Verbascum thapsus
New York fern	Dryopteris noveboracensis
Oak	Quercus spp.
Basket (or Swamp chestnut) oak	Quercus michauxii
Bear (or Scrub) oak	Quercus ilicifolia
Black oak	Quercus velutina
Blackjack oak	Quercus marilandica
Chestnut oak	Quercus prinus
Chinkapin (or Dwarf chestnut) oak	Quercus prinoides
Pin oak	Quercus palustris
Post oak	Quercus stellata
Red oak	Quercus rubra

Common Name	Genus and Species Name
Scarlet oak	Quercus coccinea
Spanish oak	Quercus falcata
Swamp chestnut (or Basket) oak	Quercus michauxii
Swamp white oak	Quercus bicolor
Water oak	Quercus nigra
White oak	Quercus alba
Willow oak	Quercus phellos
Yellow oak	Quercus muehlenbergii
Oatgrass, tall	Arrhenatherum elatius
Orache	Atriplex patula
Orchard grass	Dactylis glomerata
Orchid, white fringed	Habenaria blephariglottis
Orchid, snowy	Habenaria nivea
Palm	Palmaceae family
Panic grass	Panicum spp.
Partridge berry	Mitchella repens
Pepperbush, sweet	Clethra alnifolia
Persimmon tree	Diospyros virginiana
Pickerelweed	Pontederia spp.
Pigeon wheat moss	Polytrichum commune
Pignut (a Hickory)	Carya glabra
Pin cushion moss	Leucobryum glaucum
Pine	Pinus spp.
Loblolly pine	Pinus taeda
Pitch pine	Pinus rigida
Pond pine	Pinus serotina
Red pine	Pinus resinosa
Short leaf pine	Pinus echinata
Virginian pine	Pinus virginiana
White pine	Pinus strobus
Pinxter flower	Rhododendron nudiflorum
Pipewort, ten-angled	Eriocaulon decangulare
Pitcher plant	Sarracenia purpurea
Plantain	Plantago spp.
Plumegrass (or Reed grass)	Phragmites communis
Poison ivy	Rhus radicans
Pokeweed	Phytolacca americana
Polypody fern	Polypodium virginianum
Pondweed	Potamogeton spp.
Poplar, balsam	Populus balsamifera
Poverty-grass	Danthonia spicata

Common Name	Genus and Species Name
Prickly lettuce	Lactuca scariola
Pyxie "moss"	Pyxidanthera barbulata
Pickering morning glory	Breweria pickeringi
Queen Anne's lace (or Wild carrot)	Daucus carota
Radish, wild	Raphanus raphanistrum
Ragweed	Ambrosia spp.
Redtop	Agrostis alba
Redwood	Sequoia sempervirens
Reed grass (or Plumegrass)	Phragmites communis
Reindeer moss	Cladonia rangiferina
Rhododendron	Rhododendron spp.
Rice, wild	Zizania aquatica
Rock moss	Dicranum spp.
Rock polypody fern	Polypodium virginianum
Rock tripe	Umbilicaria spp.
Rocket, yellow	Barbarea vulgaris
Rockrose	Helianthemum spp.
Rose	Rosa spp.
Marsh rose	Rosa palustris
Multiflora rose	Rosa multiflora
Rosebay (or Great laurel)	Rhododendron maximum
Rose pogonia	Pogonia ophioglossoides
Royal fern	Osmunda regalis
Rushes	Juncus spp.
Sabal palm	Sabal palmetto
Sagebrush	Artemisia tridentata
Salt-marsh grass	Spartina spp.
Salt-marsh cordgrass (or Thatch) grass	Spartina alterniflora
Salt-meadow (or Salt-hay) grass	Spartina patens
Salt-marsh aster	Aster tenuifolius
Saltwort	Salsola kali
Samphire	Salicornia spp.
Sand myrtle	Leiophyllum buxifolium
Sand spurry, red	Spergularia spp.
Sarsaparilla, wild	Aralia nudicaulis
Sassafras	Sassafras albidum
Sawbrier	Smilax glauca
Sea blite	Suaeda spp.

Common Name	Genus and Species Name
Sea lavender	Limonium spp.
Sea myrtle	Baccharis halimifolia
Sea rocket	Cakile edentula
Sedges	Cyperaceae family (Carex – one genus)
Serviceberry (or Juneberry or Shadbush)	Amelanchier spp.
Shadscale (or Orache)	Atriplex spp.
Skunk cabbage	Symplocarpus foetidus
Snakeroot, black	Sanicula spp.
Snakeroot, white	Eupatorium rugosum
Solomon's seal	Polygonatum pubescens
Solomon's seal, false	Smilacina racemosa
Sphagnum (or Bog) moss	Sphagnum spp.
Spicebush	Lindera benzoin
Spike grass	Distichlis spp.
Spike rush	Eleocharis spp.
Spring beauty	Claytonia spp.
Spruce	Picea spp.
Black spruce	Picea mariana
St. John's-wort	Hypericum spp.
Staggerbush	Lyonia mariana
Starflower	Trientalis borealis
Steeple-bush	Spiraea tomentosa
Strangler-fig	Ficus aurea
Strawberry bush	Euonymus americanus
Strawberry, wild	Fragaria virginiana
Sumac	Rhus spp.
Dwarf (or Winged) sumac	Rhus copallina
Smooth sumac	Rhus glabra
Staghorn sumac	Rhus typhina
Sundew	Drosera spp.
Swamp-pink	Helonias bullata
Sweetbay (or Magnolia)	Magnolia virginiana
Sweetfern	Comptonia peregrina
Sweetflag	Acorus calamus
Sweetgum	Liquidambar styraciflua
Sycamore	Platanus occidentalis
Tamarack (or Larch)	Larix lariciana
Tamarind, wild	Lysiloma spp.
Tanbark-oak	Lithocarpus densiflora

Common Name	Genus and Species Name
Thatch grass (or Salt-marsh cord-grass)	Spartina alterniflora
Thistle, Canada	Cirsium arvense
Thread moss	Bryum spp.
Tick trefoil	Desmodium spp.
Timothy	Phleum pratense
Touch-me-not (or Jewelweed)	Impatiens spp.
Trailing arbutus	Epigaea repens
Tree-of-heaven	Ailanthus altissima
Trout-lily (or Dog's-tooth violet)	Erythronium americanum
Tulip tree (or Yellow poplar)	Liriodendron tulipifera
Turkeybeard	Xerophyllum asphodeloides
Viburnum	Viburnum spp.
Maple-leaved viburnum	Virburnum acerifolium
Violet	Viola spp.
Virginia chain fern	Woodwardia virginica
Virginia creeper	Parthenocissus quinquefolia
Walnut, black	Juglans nigra
Water dock	Rumex verticillatus
Water lily	Nymphaea spp.
Water lily, fragrant	Nymphaea odorata
Water willow (or Swamp-loose-strife)	Decodon verticillatus
Wax myrtle	Myrica heterophylla
Willow	Salix spp.
Winterberry (or Black alder)	Ilex verticillata
Winterberry, smooth	Ilex laevigata
Wintercress (or Yellow rocket)	Barbarea vulgaris
Wintergreen, aromatic	Gaultheria procumbens
Wintergreen, spotted	Chimaphila maculata
Witch hazel	Hamamelis virginiana
Wormwood	Artemisia spp.
Yarrow (or Milfoil)	Achillea millefolium
Yellow poplar (or Tulip tree)	Liriodendron tulipifera
Yellow rocket (or wintercress)	Barbarea vulgaris
Yellow-eyed grass	Xyris spp.
Yellow-eyed grass, Carolinian	Xyris caroliniana

The following plants of the heath (Ericaceae) family grow in New Jersey:

Common Name	Genus Name
Andromeda (or Bog rosemary)	Andromeda
Bearberry	Arctostaphylos
Leatherleaf	Chamaedaphne
Trailing arbutus	Epigaea
Aromatic wintergreen	Gaultheria
Black huckleberry, blue huckleberry, and dangleberry	Gaylussacia
Laurel, mountain and sheep	Kalmia
Labrador tea	Ledum
Sand myrtle	Leiophyllum
Fetterbush	Leucothoe
Maleberry and staggerbush	Lyonia
Rhododendron (Rosebay), Pinxter flower, Swamp azalea	Rhododendron
Cranberry, deerberry, and several species of blueberry	Vaccinium

Appendix III, Part B —
Cross Reference, Plant Scientific Names to Plant Common Names

Genus and Species Name	Common Name
Abies spp.	Fir
Acer	Maple
negundo	Ash-leaved maple (or box elder)
pensylvanicum	Striped maple
platanoides	Norway maple
rubrum	Red maple
rubrum var. trilobum	Red maple, three-lobed variety
saccharinum	Silver maple
saccharum	Sugar maple
Achillea millefolium	Milfoil (or Yarrow)
Acorus calamus	Sweetflag
Actaea spp.	Baneberry
Adenostoma fasciculatum	Chamiso
Agrostis	Bent grass
alba	Redtop
Ailanthus altissima	Tree-of-heaven
Allium canadense	Wild garlic
Alnus	Alder
rugosa	Alder, speckled
serrulata	Alder, common
Ambrosia spp.	Ragweed
Amelanchier spp.	Serviceberry (or Juneberry or Shadbush)
Ammophila breviligulata	Dunegrass (Marram or Beachgrass)
Amphicarpa spp.	Hog peanut
Anemonella thalictroides	Anemone, rue
Andromeda glaucophylla	Bog rosemary

324

Genus and Species Name	Common Name
Andropogon	Beardgrass
scoparius	Little bluestem grass
virginicus	Broom sedge
Anthemis arvensis	Corn chamomile
Aquilegia canadensis	Columbine
Aralia nudicaulis	Sarsaparilla, wild
Arbutus menziesii	Madrone
Arctostaphylos tomentosa	Manzanita
Arctostaphylos uva-ursi	Bearberry
Arethusa bulbosa	Arethusa
Arisaema triphyllum	Jack-in-the-pulpit
Arrhenatherum elatius	Oatgrass, tall
Artemisia spp.	Wormwood
Artemisia stelleriana	Dusty miller
Artemisia tridentata	Sagebrush
Asarum canadense	Ginger, wild
Asclepias spp.	Milkweed
Aster	Aster
divaricatus	Woodland aster
linariifolius	Stiff-leaved aster
nemoralis	Bog aster
tenuifolius	Salt-marsh aster
Atriplex patula	Orache
Baccharis halimifolia	Sea myrtle
Baptisia tinctoria	Indigo, wild
Barbarea vulgaris	Rocket, yellow (or wintercress)
Berberis spp.	Barberry
Betula	Birch
lenta	Sweet (or Black) birch
lutea	Yellow birch
nigra	River birch
papyrifera	Paper (or White) birch
populifolia	Gray birch
Biatorella clavis	Brownie's buttons
Breweria pickeringi	Pickering morning glory
Bromus spp.	Brome grass
Bryum spp.	Thread moss
Bursera spp.	Gumbo limbo
Cabomba caroliniana	Fanwort
Cactaceae family	Cactus plants

Genus and Species Name	Common Name
Cakile edentula	Sea rocket
Calapogon pulchellus	Grass-pink
Carex	Sedge
Carpinus caroliniana	Hornbeam, American (or Iron-wood)
Carya	Hickory
glabra	Pignut (a Hickory)
ovalis	Red hickory
ovata	Shagbark hickory
tomentosa	Mockernut (a Hickory)
Castanea dentata	Chestnut, American
Castanopsis chrysophylla	Chinquapin, giant
Celastrus scandens	Bittersweet, climbing
Celtis occidentalis	Hackberry
Cephalanthus occidentalis	Buttonbush
Cerastium vulgatum	Mouse-eared chickweed
Chamaecyparis thyoides	Cedar, Southern white
Chamaedaphne calyculata	Leatherleaf
Chimaphila maculata	Wintergreen, spotted
Chrysanthemum leucanthemum	Daisy, white (or Oxeye daisy)
Cichorium intybus	Chicory
Circaea spp.	Enchanter's nightshade
Circium arvense	Thistle, Canada
Cladonia cristatella	British soldier
Cladonia rangiferina	Reindeer moss
Claytonia spp.	Spring beauty
Clethra alnifolia	Pepperbush, sweet
Comptonia peregrina	Sweetfern
Corema conradii	Broom crowberry
Cornus	Dogwood
alternifolia	Pagoda (or Alternate-leaved) dog-wood
amomum	Silky dogwood
florida	Flowering dogwood
racemosa	Gray dogwood
stolonifera	Red dogwood
Corylus	Hazel (or Hazelnut)
americana	American hazel
cornuta	Beaked hazel
Crataegus spp.	Hawthorn
Cyperaceae family	Sedges

Genus and Species Name	Common Name
Cypripedium spp.	Lady's slipper
acaule	Stemless lady's slipper
Dactylis glomerata	Orchard grass
Danthonia spicata	Poverty grass
Daucus carota	Queen Anne's lace (or Wild carrot)
Decodon verticillatus	Loosestrife, swamp (or Water willow)
Deschampsia flexuosa	Hairgrass
Desmodium spp.	Tick trefoil
Dicentra cucullaria	Dutchman's breeches
Dicranum spp.	Rock moss
Digitaria spp.	Crabgrass
Diospyros virginiana	Persimmon tree
Distichlis spp.	Spike grass
Drosera spp.	Sundew
Dryopteris	Shield (or Wood) fern
marginalis	Marginal shield fern
noveboracensis	New York fern
thelypteris	Marsh fern
Eleocharis spp.	Spike rush
Epifagus virginiana	Beech drops
Epigaea repens	Trailing arbutus
Ericaceae family	Heath plants (see Footnote, Appendix, Part A)
Erigeron canadensis	Horse-weed
Eriocaulon decangulare	Pipewort, ten-angled
Eriophorum virginicum	Cotton grass, tawny
Erythronium americanum	Trout-lily (or Dog's-tooth violet)
Euonymus americanus	Strawberry bush
Eupatorium	Thoroughwort
perfoliatum	Boneset
purpureum	Joe-Pye-weed
rugosum	Sankeroot, white
Euphorbia ipecacuanhae	Ipecac, wild
Fagus grandifolia	Beech
Ficus aurea	Strangler-fig
Fragaria virginiana	Strawberry, wild
Franseria spp.	Bur sage

Genus and Species Name	Common Name
Fraxinus	Ash
americana	White ash
pennsylvanica	Green ash
Gaultheria procumbens	Wintergreen, aromatic
Gaylussacia	Huckleberry
baccata	Black huckleberry
frondosa	Dangleberry (or Blue huckleberry)
Gentiana spp.	Gentian
autumnalis	Pine Barren gentian
Geranium maculatum	Cranesbill
Habenaria blephariglottis	Orchid, white-fringed
Habenaria nivea	Orchid, snowy
Hamamelis virginiana	Witch hazel
Helianthemum spp.	Rockrose
canadense	Frostweed
Helonias bullata	Swamp-pink
Hepatica americana	Hepatica
Hibiscus palustris	Marsh mallow (or Mallow rose)
Hieracium	Hawkweed
pratense	King Devil hawkweed
Hudsonia	Hudsonia
ericoides	Heather, false (or Golden heather)
tomentosa	Beach heather (or Heath)
Hypericum spp.	St. John's-wort
Ilex	Holly
glabra	Inkberry (or Gallberry)
laevigata	Winterberry, smooth
opaca	Holly, American
verticillata	Alder, black (or Winterberry)
Impatiens spp.	Jewelweed (or Touch-me-not)
Iris versicolor	Blue flag
Iva frutescens	Marsh elder
Juglans nigra	Black walnut
Juncus	Rushes
gerardi	Black marsh (or Black) grass
Juniperus virginiana	Cedar, red

Genus and Species Name	Common Name
Kalmia	Laurel
angustifolia	Sheep laurel
latifolia	Mountain laurel
Lactuca scariola	Prickly lettuce
Larix laricina	Larch, American (or Tamarack)
Larrea spp.	Creosote bush
Lathyrus maritimus	Beach pea
Ledum groenlandicum	Labrador tea
Leiophyllum buxifolium	Sand myrtle
Lemna minor	Duckweed
Lespedeza spp.	Bush clover
Leucobryum glaucum	Pin cushion moss
Leucothoe racemosa	Fetterbush
Liatris graminifolia	Blazing star
Limonium spp.	Sea lavender
Linaria vulgaris	Butter-and-eggs
Lindera benzoin	Spicebush
Liquidambar styraciflua	Sweetgum
Liriodendron tulipifera	Tulip tree (or Yellow poplar)
Lithocarpus densiflora	Tanbark-oak
Lobelia	Lobelia
cardinalis	Cardinal flower
Lonicera	Honeysuckle
japonica	Japanese honeysuckle
Lophiola americana	Golden-crest
Lycopodium	Club moss
carolinianum	Club moss, Carolinian
complanatum	Ground cedar
obscurum	Ground pine
Lygodium palmatum	Climbing fern
Lyonia ligustrina	Maleberry
Lyonia mariana	Staggerbush
Lysiloma spp.	Tamarind, wild
Lysimachia quadrifolia	Loosestrife, whorled
Magnolia virginiana	Sweetbay (or Magnolia)
Maianthemum canadense	Lily-of-the-valley, false
Melampyrum lineare	Cow-wheat
Menispermum canadense	Moonseed
Mertensia virginica	Bluebells (or Virginia cowslip)

Genus and Species Name	Common Name
Mitchella repens	Partridge berry
Myrica heterophylla	Wax myrtle
Myrica pensylvanica	Bayberry
Narthecium americanum	Asphodel, bog
Nymphaea	Water lily
odorata	Water lily, fragrant
Nyssa sylvatica	Gum, black (or Sour gum)
Oenothera biennis	Evening-primrose
Orontium aquaticum	Golden club
Osmunda	Flowering fern
cinnamomea	Cinnamon fern
regalis	Royal fern
Ostrya virginiana	Hop hornbeam
Palmaceae family	Palm
Panicum	Panic grass
depauperatum	Impoverished grass
Parthenocissus quinquefolia	Virginia creeper
Peltandra virginica	Arrow-arum
Phleum pratense	Timothy
Phoradendron flavescens	Mistletoe, American
Phragmites communis	Reed (or plume) grass
Phytolacca americana	Pokeweed
Picea	Spruce
mariana	Black spruce
Pinus	Pine
echinata	Short leaf pine
resinosa	Red pine
rigida	Pitch pine
serotina	Pond pine
strobus	White pine
taeda	Loblolly pine
virginiana	Virginian pine
Plantago spp.	Plantain
Platanus occidentalis	Sycamore
Pluchea spp.	Marsh fleabane
Poa spp.	Blue grass
Podophyllum peltatum	May apple
Pogonia ophioglossoides	Rose pogonia

Genus and Species Name	Common Name
Polygala	Milkwort
lutea	Milkwort, orange
Polygonatum pubescens	Solomon's seal
Polypodium virginianum	Rock polypody fern
Polystichum acrostichoides	Christmas fern
Polytrichum commune	Pigeon wheat moss
Polytrichum juniperinum	Haircap moss
Pontederia spp.	Pickerelweed
Populus	Aspen (or Poplar)
balsamifera	Poplar, balsam
grandidentata	Large-toothed aspen
tremuloides	Trembling (or Quaking) aspen
Potamogeton spp.	Pondweed
Potentilla spp.	Cinquefoil
Prunus	Cherry
avium	Sweet cherry
maritima	Beach plum
serotina	Black cherry
Pteridium aquilinum	Bracken fern
Pyrus arbutifolia	Chokeberry, red
Pyrus malus	Apple
Pyrus melanocarpa	Chokeberry, black
Pyxidanthera barbulata	Pyxie "moss"
Quercus	Oak
alba	White oak
bicolor	Swamp white oak
coccinea	Scarlet oak
falcata	Spanish oak
ilicifolia	Bear (or Scrub) oak
marilandica	Blackjack oak
michauxii	Basket (or Swamp chestnut) oak
muehlenbergii	Yellow oak
nigra	Water oak
palustris	Pin oak
phellos	Willow oak
prinoides	Chinkapin (or Dwarf-chestnut) oak
prinus	Chestnut oak
rubra	Red oak
stellata	Post oak
velutina	Black oak

Genus and Species Name	Common Name
Raphanus raphanistrum	Wild radish
Rhizophora mangle	Mangrove, red
Rhododendron	Rhododendron
maximum	Rosebay (or Great laurel)
nudiflorum	Pinxter flower
viscosum	Axalea, swamp (or Swamp honeysuckle)
Rhus	Sumac
copallina	Dwarf (or Winged) sumac
glabra	Smooth sumac
radicans	Poison ivy
typhina	Staghorn sumac
Ribes spp.	Gooseberry
Robinia pseudo-acacia	Locust, black
Rosa	Rose
multiflora	Multiflora rose
palustris	Marsh rose
Rubus	Bramble
allegheniensis	Blackberry
flagellaris	Dewberry
Rudbeckia serotina	Black-eyed Susan
Rumex verticillatus	Water dock
Sabal palmetto	Sabal palm
Sagittaria latifolia	Arrow-head
Salicornia spp.	Glasswort (or Samphire)
Salix spp.	Willow
Salsola kali	Saltwort
Sambucus canadensis	Elderberry
Sanicula spp.	Snakeroot, black
Sarracenia purpurea	Pitcher plant
Sassafras albidum	Sassafras
Schizaea pusilla	Curly-grass fern
Scirpus spp.	Bulrush
Sequoia sempervirens	Redwood
Sericocarpus asteroides	White-topped aster
Setaria spp.	Foxtail grass
Smilacina racemosa	Solomon's seal, false
Smilax	Greenbrier (or Catbrier)
glauca	Sawbrier
Solidago	Goldenrod
bicolor	White goldenrod

Genus and Species Name	Common Name
caesia	Blue-stemmed goldenrod
sempervirens	Seaside goldenrod
Sorghastrum nutans	Indian grass
Sparganium spp.	Bur reed
Spartina	Salt-marsh grass
alterniflora	Salt-marsh cordgrass (or Thatch) grass
patens	Salt-meadow (or Salt-hay) grass
Spergularia spp.	Sand spurry, red
Sphagnum spp.	Sphagnum (or Bog) moss
Spiraea tomentosa	Steeple-bush
Stellaria spp.	Chickweed
Stipa avenacea	Black oat grass
Suaeda spp.	Sea blite
Symplocarpus foetidus	Skunk cabbage
Taraxacum officinale	Dandelion
Tephrosia virginiana	Goat's-rue
Tilia americana	Basswood
Trientalis borealis	Starflower
Tsuga canadensis	Hemlock
Typha spp.	Cattail
Ulmus	Elm
americana	American elm
rubra	Red (or Slippery) elm
Umbellularia californica	California laurel
Umbilicaria spp.	Rock tripe
Utricularia spp.	Bladderwort
Uvularia spp.	Bellwort
Vaccinium	Blueberry
atrococcum	Black highbush blueberry
corymbosum	Highbush blueberry
macrocarpon	Cranberry
stamineum	Deerberry
vacillans	Lowbush blueberry
Vallisneria americana	Celery, wild
Verbascum thapsus	Mullein, common
Viburnum	Viburnum
acerifolium	Maple-leaved viburnum
dentatum	Arrowwood
prunifolium	Black haw

Genus and Species Name	Common Name
Vicia villosa	Hairy-vetch
Viola spp.	Violet
Vitis	Grape
aestivalis	Summer grape
novae-angliae	New England grape
Woodwardia	Chain fern
virginica	Virginia chain fern
Xerophyllum asphodeloides	Turkeybeard
Xyris	Yellow-eyed grass
caroliniana	Yellow-eyed grass, carolinian
Zizania aquatica	Rice, wild
Zostera mariana	Eelgrass

Index

Acidity: bogs, 149; plant habitats and, 104
Agriculture, 3
Air mass movements, 56
Air pollution, vegetation and, 12–13, 63, 79–80, 275, 277. *See also* Pollution
Allaire State Park, 298, 303
Animals: as an ecosystem component, 14–15; plant relationships and, 89–94, 127–128
Arney's Mount, 45, 46

Barnegat Bay, 238
Barrens. *See* Pine Barrens
Barrier Islands, 235–238
Bass River State Park, 45, 295
Beach Heather community, 243–244
Beacon Hill, 45
Beavers, impact on drainage and vegetation, 93, 150, 296
Bedrock, geologic map, 48–49
Beech forest, 210, 302
Beech-Maple forest region, 261
Belle Plain State Forest, 159, 298
Biological interrelationships, 14–15, 82–94
Biotic community, 98
Bog: defined, 131–132; development of, 131–134; habitats, 136; location, 105, 132–133; man and, 143–144; vegetation of, 131–145, 294–295
Boreal forest. *See* Northern Conifer forest formation

Brigantine National Wild Life Refuge, 114
Budd Lake, 37
Bull's Island, 153, 295

Cape May, 158, 211, 238, 298
Cattail Marsh, 123, 124, 292–293
Cedar bogs, 138–142. *See also* Bog
Cenozoic era, 30, 35–38; division of, 35–36; Tertiary time, 35
Chaparral formation, 258
Cheesequake State Park, 114, 212, 291–292, 303
Chestnut blight, 87–88, 167
Chestnut Oak forest type, 189–193, 301
Climate: as an ecosystem component, 11–13, 54–64; changes in, 86; continental nature of, 54–55; description of, 54–64; frostfree periods, 59; growing season periods, 58; daylight, 60, 61–63; plant habitats and, 102; precipitation, 61; temperature, 55–60; variations, 60; relation to vegetation, 11–12
Coastal features, 234–241
Coastal Plains: formation of, 35, 44–47; geological division, 207–208; soils, 49–52, 99, 208. *See also* South Jersey
Climax vegetation, defined, 24, 108
Coasts, formation of, 234–241
Community, biotic and plant, 98
Continental climate, defined, 55
Controlled burning, 230–231, 304

335